BIOLOGY
Visualizing Life

LABORATORY EXPERIMENTS
TEACHER'S EDITION

HOLT, RINEHART AND WINSTON
Austin · *New York* · *Orlando* · *Chicago* · *Atlanta* · *San Francisco* · *Boston* · *Dallas* · *Toronto* · *London*

Reviewers:

Don Chmielowiec
Alex Molinich
George Nassis
Laboratory Investigations
WARD'S Natural Science Establishment, Inc.
Rochester, New York

Kenneth Rainis
Safety
WARD'S Natural Science Establishment, Inc.
Rochester, New York

Cover design: Foca, Inc., New York, NY
Front cover: (fly) © Herman Eisenbeiss, Photo Researchers
Back cover: (chromosomes) © Biophoto Associates, Photo Researchers
               (dinosaur) Mark Hallet, Mark Hallet Illustrations
               (DNA) R Margulies, Virginia Ferrante, Margulies Medical Associates
               (eagle) © Jim Simmen, AllStock
               (Earth) © Frank Rossotto, The Stock Market
               (pea flowers) Karen Kluglein, rep. Mattleson Associates, Ltd.
               (pea pod, sunflower) Sergio Purtell, Foca, Inc.

For permission to reprint copyrighted material, grateful acknowledgment is made to the following sources:

National Association of Biology Teachers, Reston, VA 22090: From "NABT Guidelines for the Use of Live Animals." Copyright © 1991 by National Association of Biology Teachers.

St. Martin's Press, Inc. and Harrow House Editions Limited: Adapted from illustrations from *After Man: A Zoology of the Future* by Dougal Dixon, illustrated by Diz Wallis and John Butler. Copyright © 1981 by Harrow House Editions Limited. Published by St. Martin's Press, Inc., 1981.

Printed in the United States of America

ISBN   0-03-076416-5

1 2 3 4 5 6 7 8 9   239   96 95 94 93

Contents

FOCUS ACTIVITIES

Chapter

INVESTIGATIONS

Complete List of *Biology: Visualizing Life* Lab Experiments

	Biology: Visualizing Life PE and ATE	Lab Manual

Using the Focus Activities

Each of the 34 Focus Activities provides a hands-on experience that will introduce students to one or more of the fundamental concepts needed to master the chapter content. Students need not be familiar with the chapter content before beginning the Focus Activity. Each activity requires few materials and can usually be performed within one class period. Most of the activities do not require a laboratory setting and can be done in any classroom.

The Focus Activities will help students to:

- move from the concrete to the abstract
- prepare to master the chapter content
- see the relevance of the chapter content to everyday life

Each two-page Focus Activity has the following parts:

Background orients the students, providing them with the introductory information necessary to perform the activity and to understand its connection to the chapter content.

Objectives identify the science process skills that students will be expected to demonstrate during the investigation.

Materials lists the supplies needed for the activity.

Analysis asks students to think critically about the results of the activity. Students are asked questions that will require them to compare observations, analyze data, evaluate methods, make inferences, and apply concepts.

Although each Focus Activity is designed to precede the corresponding chapter of *Biology: Visualizing Life*, the activities can also be used as additional lab activities after students have completed the chapter, or at any point during a chapter lesson.

Using Gowin's Vee to Link Lecture to Laboratory

Gowin's Vee is a scaffold device that illustrates why knowledge is the end result of the process of inquiry. You can use the Gowin's Vee to make the connection between lecture and laboratory, to guide science fair research, and to make oral reports and lab reports easier to understand.

Using the Vee for Lab Reporting

Vee report sheets are included with all of the laboratory investigations in this manual. You can also use the Vee with the investigations in the textbook. Students fill out the left side and center portions of the Vee as a pre-lab activity. The right side of the Vee is completed after the investigation is completed. The completed Vee serves as a comprehensive assessment of students' understandings and findings from the Investigation.

   The diagram below shows the common elements of the Vee. This Vee summarizes an Investigation concerning osmosis. Students will complete the parts of the Vee in the order described on the following page.

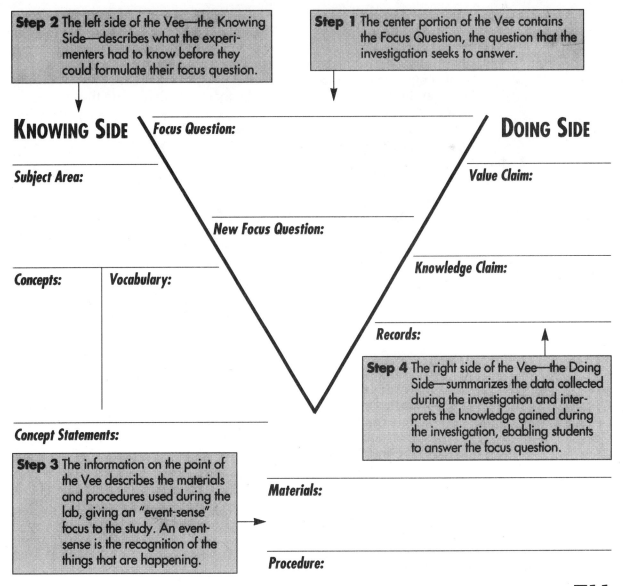

Step 2 The left side of the Vee—the Knowing Side—describes what the experimenters had to know before they could formulate their focus question.

Step 1 The center portion of the Vee contains the Focus Question, the question that the investigation seeks to answer.

KNOWING SIDE

Focus Question:

DOING SIDE

Subject Area:

Value Claim:

New Focus Question:

Concepts: Vocabulary:

Knowledge Claim:

Records:

Step 4 The right side of the Vee—the Doing Side—summarizes the data collected during the investigation and interprets the knowledge gained during the investigation, ebabling students to answer the focus question.

Concept Statements:

Step 3 The information on the point of the Vee describes the materials and procedures used during the lab, giving an "event-sense" focus to the study. An event-sense is the recognition of the things that are happening.

Materials:

Procedure:

Before the Lab

Students write a Focus Question (Step 1) and fill in the Knowing Side of the Vee (Step 2). Then they complete the Materials and Procedure sections of the Vee (Step 3).

Step 1:

The **Focus Question** describes the objects and main events of the investigation and indicates the kinds of records that will be collected. This question can usually be constructed from reading the Introduction, Materials, and the Procedure sections of the investigation. The focus question is the main question that the experiment seeks to answer.

Step 2:

Subject Area identifies one or more broad, comprehensive areas covered by the investigation. The subject area encompasses many of the concepts involved in a lab and explains why a phenomenon occurs.

Concepts and **Vocabulary** Concepts are the ideas that delineate the scope of the investigation. Vocabulary includes any new terms used in the investigation.

Concept Statements demonstrate one's knowledge and understanding of the fundamental ideas on which the investigation is based. Concept statements relate two or more concepts to each other.

Step 3:

Materials and **Procedure Materials** are the objects that enable one to perform the investigation. Procedure summarizes the processes that occurred to collect data during the investigation. These two sections help students to acknowledge the key events that occurred during a lab, and to recall and recognize the materials that made these outcomes possible.

After the Lab

Students complete the Doing Side of the Vee (Step 4).

Step 4:

Records are the data collected during the investigation. They should be presented in an organized fashion, such as tables, charts, and graphs. Records that cannot fit on the front of the Vee can be placed on the back of the Vee under the heading "Additional Records and Observations."

Knowledge Claims are statements or claims about what students learned during the lab. Knowledge claims answer the focus question. They may also lead to new focus questions for new investigations.

Value Claim describes the significance of the knowledge gained in the investigation.

New Focus Question describes the next possible stage in the investigation. Formulating a new focus question helps students to see that new scientific knowledge is based on information gained in earlier studies. Students can see how the knowledge claims produced during this investigation could become the concept statements for the new study proposed in the new focus question.

Teaching Students to Use the Vee

As a summary of a laboratory experience, a teacher can show how the laboratory experiment can be dissected into components to fit on the Vee. In future laboratory experiments, the teacher can help the students develop the left side and central parts of the Vee. At this time, the point can be made that the left side is a summary of non-lab/class/lecture discussions. By the time they construct their second Vee, students understand how to construct it and seem to be more adept at identifying its various components. Because the Subject Area portion of the Vee seems to be the most difficult for students, one or more Subject Areas have already been identified on the Vee forms that accompany each lab.

Since students tend to work in lab groups or with lab partners, the Vee can be constructed cooperatively. If you feel it is more valuable to the student's own understanding that the Vee be constructed individually, that is also an option. If you give the students your criteria of grading, they can actually "grade" their own Vee form reports before handing them in. This will enable them to check their own work.

One of the most valuable results of using Gowin's Vee is that students come to know the tentativeness of scientific truths. They see that knowledge claims are indeed claims and not conclusions.

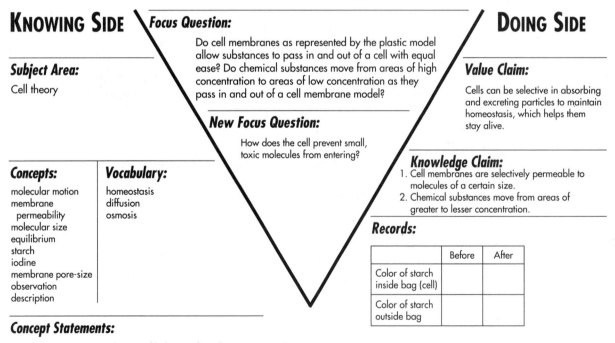

KNOWING SIDE

Focus Question:

Do cell membranes as represented by the plastic model allow substances to pass in and out of a cell with equal ease? Do chemical substances move from areas of high concentration to areas of low concentration as they pass in and out of a cell membrane model?

New Focus Question:

How does the cell prevent small, toxic molecules from entering?

Subject Area:

Cell theory

Concepts:

molecular motion
membrane
  permeability
molecular size
equilibrium
starch
iodine
membrane pore-size
observation
description

Vocabulary:

homeostasis
diffusion
osmosis

DOING SIDE

Value Claim:

Cells can be selective in absorbing and excreting particles to maintain homeostasis, which helps them stay alive.

Knowledge Claim:
1. Cell membranes are selectively permeable to molecules of a certain size.
2. Chemical substances move from areas of greater to lesser concentration.

Records:

	Before	After
Color of starch inside bag (cell)		
Color of starch outside bag		

Concept Statements:

1. Homeostasis, or a steady state of balance, of a cell depends on movement of materials in and out of a cell.
2. Molecules constantly move at random from areas of greater to areas of lesser concentration until they reach equilibrium.
3. Osmosis is the diffusion of water.
4. Cells have membranes.
5. The cell is the basic unit of structure and function of living things.

Materials:

starch solution, iodine solution, membrane, beaker, plastic bag, test tube

Procedure:

Starch and iodine solutions on either side of a model membrane are observed within a beaker, test tube, and plastic bag.

Assessing Student-Constructed Vee Reports

You can use the comprehensive criteria shown below to assess each part of a student-constructed Vee. The higher the score, the more complete and correct the Vee part.

Focus Question

0 No question is identified.

1 A question is identified but does not focus on the objects and the major event or on the conceptual side of the Vee.

2 A focus question is identified; includes concepts, but does not suggest materials or procedure, or the wrong materials and procedures are identified in relation to the rest of the laboratory exercise.

3 A clear focus question is identified; includes concepts to be used and suggests the procedure and materials.

Materials/Procedure

0 No materials and procedures are identified.

1 The materials and procedure are identified but are inconsistent with the focus question.

2 The materials and procedure are identified and are consistent with the focus question.

3 Same as above, but also suggests what records will be taken.

Concepts/Concept Statements

0 No conceptual side is identified.

1 A few concepts are identified, and no concept statements, or a concept written is really the knowledge claim sought in the laboratory exercise.

2 Concepts and at least one type of concept statement are identified.

3 Concepts and at least two types of concept statements are identified.

4 Concepts, two types of concept statements, and all relevant vocabulary are identified.

Records

0 No records or observations are identified.

1 Records are identified but are inconsistent with the focus question or the major event.

2 Records or observations are identified, but not both.

3 Records are identified for the major event, but observations are inconsistent with the intent of the focus question.

4 Records are identified for the major event, and observations are consistent with the focus question and the grade level and ability of the student.

Knowledge Claim

0 No knowledge claim is identified.

1 A knowledge claim is unrelated to the left-hand side of the Vee.

2 A knowledge claim that includes a concept that is used in an improper context, or any generalization that is inconsistent with the records.

3 A knowledge claim is made that includes the concepts from the focus question and is derived from the records.

4 Same as above, but the knowledge claim leads to a new focus question.

New Focus Question

0 No new focus question is given.

1 A new focus question consistent with the knowledge claim is identified.

Value Claim

0 No new value claim is given.

1 A claim is given consistent with the significance of the investigation describing the usefulness of the knowledge claim for pure or applied scientific endeavors.

Maintaining a Safe Lab

Building Safety Partnerships: You're Not Alone

As a teacher, you will be held accountable by students, parents, and administrators for a safe laboratory. You will also be responsible for maintaining an environment that is ethically appropriate and conducive to learning. These goals are related. An unsafe laboratory is likely to be one that is not learning-oriented, and one in which irresponsibility and unethical treatment of living things (and lab partners) run rampant.

A safe laboratory can be achieved only through a partnership between all parties concerned, not one involving only students or teachers. Teachers must actively boost "safety-consciousness" among students, fellow faculty members, administrators, and parents. For success, everyone must agree to respect the same laboratory rules, to obtain and use the proper safety equipment, and to take the appropriate precautions during a lab investigation.

A small investment of time and effort can have a big payoff if it prevents an accident. With a little planning and thoughtfulness beforehand, accidents that do occur will not be as serious.

An excellent way to start building this safety partnership with your students is the "safety contract" in the pupil's edition of the lab experiments. Have each student fill one out and return it to you. Keep them on file, in case you need to remind students of their promises.

Where to Start

In each lab investigation, safety symbols are included and specific safety procedures are highlighted where appropriate. In addition, the beginning of the pupil's edition of the lab investigations (found in the Teacher's Resource Binder) also includes detailed descriptions of each safety symbol and hazard, and the precautions related to each one. The safety symbol descriptions are included in expanded form in this section.

The information included in this section will help you plan and maintain a safe and healthy laboratory environment.

- **This information is not all-inclusive.** Each school's lab situation is different, and no publication could list safe practices for all situations that could possibly arise.

- **This information probably does not include all the legal requirements you will need to follow.** Be sure that you are aware of any federal, state, or local laws that may cover your lab. Many, but not all, laboratory safety regulations are made and enforced by the Occupational Safety and Health Administration (OSHA) or by an OSHA-approved state safety agency. Teachers should keep in mind that OSHA's regulations were primarily designed for private industry, but some states have extended the regulations to cover public schools as well.

Although laws and regulations can vary from place to place and from time to time, you can build a safe program suited to your situation using the information presented here.

Safety Symbols

Eye Safety

- **Wear approved chemical safety goggles as directed.** Goggles should always be worn whenever you are working with chemicals or solutions, heating substances, using mechanical devices, or observing physical processes.

- **In case of eye contact:**
  Go to an eyewash station and flush eyes (including under the eyelids) with running water for at least 15 minutes. Notify your teacher or other adult in charge.

- **Wearing of contact lenses for cosmetic reasons is prohibited in the laboratory.** Liquids or gases can be drawn up under the contact lens and onto the eyeball. If you must wear contacts prescribed by a physician, inform your teacher. You must wear approved eyecup safety goggles.

- **Never look directly at the sun through any optical device or lens system, nor gather direct sunlight to illuminate a microscope.** Such actions will concentrate light rays that will severely burn your retina, possibly causing blindness!

Electrical Supply

- **Never use equipment with frayed cords.**

- **Assure that electrical cords are taped to work surfaces.** This will prevent falls, and equipment can't be pulled off tables.

- **Never use electrical equipment around water, nor with wet hands or clothing.**

Clothing Protection

- **Wear an apron or laboratory coat when working in the laboratory, to prevent chemicals or chemical solutions from contacting skin or contaminating street clothes.** Confine all loose clothing and long jewelry.

Animal Care

- **Do not touch or approach any animal in the wild.** Be aware of poisonous or dangerous animals in any area where you will be doing outside fieldwork.

- **Always obtain your teacher's permission before bringing any animal (or pet) into the school building.**

- **Handle any animal only as your teacher directs.** Mishandling or abuse of any animal will not be tolerated!

Sharp Object Safety

- **Use extreme care with all sharp instruments such as scalpels, sharp probes, and knives.**

- **Never use double-edged razors in the laboratory.**

- **Never cut objects while holding them in your hand.** Place objects on a suitable work surface.

Chemical Safety

- **Always wear appropriate personal protective equipment.** Eye goggles, gloves, and apron or lab coat should always be worn when working with any chemical or chemical solution.

- **Never taste, touch, or smell any substance, nor bring it close to your eyes, unless specifically told to do so by your teacher.** If you are directed by your teacher to note the odor of a substance, do so by waving the fumes towards you with your hand. Never pipet any substance by mouth; use a suction bulb as directed by your teacher.

- **Always handle any chemical or chemical solution with care.** Check the label on the bottle and observe safe-use procedures. Never return unused chemicals or solutions to their containers. Return unused reagent bottles or containers to your teacher. Store chemicals according to your teacher's directions.

- **Never mix any chemical** unless specifically told to do so by your teacher.

- **Never pour water into a strong acid or base.** The mixture can produce heat and splatter. Remember this rhyme...

  Do as you oughta—
  Add acid (or base) to water.

- **Report any spill immediately to your teacher.** Handle spills only as your teacher directs.

- **Check for the presence of any source of flames, sparks, or heat (open flame, electric heating coils, etc.) before working with flammable liquids or gases.**

Plant Safety

- **Do not ingest any plant part used in the laboratory (especially seeds sold commercially).** Do not rub any sap or plant juice on your eyes, skin, or mucous membranes.

- **Wear protective gloves (disposable polyethylene gloves) when handling any wild plant.**

- **Wash hands thoroughly after handling any plant or plant part (particularly seeds).** Avoid touching hands to your face and eyes.

- **Do not inhale or expose yourself to the smoke of any burning plant.**

- **Do not pick wildflowers or other plants unless directed to do so by your teacher.**

Proper Waste Disposal

- **Clean and decontaminate all work surfaces and personal protective equipment as directed by your teacher.**

- **Dispose of all sharps (broken glass and other contaminated sharp objects) and other contaminated materials (biological/chemical) in special containers as directed by your teacher.**

Hygienic Care

- **Keep your hands away from your face and mouth.**
- **Wash your hands thoroughly before leaving the laboratory.**
- **Remove contaminated clothing immediately; launder contaminated clothing separately.**
- **Use the proper technique demonstrated by your teacher when handling bacteria or similar microorganisms.** Examine microorganism cultures (such as petri dishes) without opening them.
- **Return all stock and experimental cultures to your teacher for proper disposal.**

Heating Safety

- **When heating chemicals or reagents in a test tube, never point the test tube toward anyone.**
- **Use hot plates, not open flames.** Be sure hot plates have an "On-Off" switch and indicator light. Never leave hot plates unattended, even for a minute. Never use alcohol lamps.
- **Know the location of laboratory fire extinguishers and fire blankets.** Have ice readily available in case of burns or scalds.
- **Use tongs or appropriate insulated holders when heating objects.** Heated objects often do not have the appearance of being hot. Never pick up an object with your hand unless you are certain it is cold.
- **Keep combustibles away from heat and other ignition sources.**

Hand Safety

- **Never cut objects while holding them in your hand.**
- **Wear protective gloves when working with stains, chemicals, chemical solutions, or with wild (unknown) plants.**

Glassware Safety

- **Inspect glassware before use; never use chipped or cracked glassware.** Use borosilicate glass for heating.
- **Do not attempt to insert glass tubing into a rubber stopper without specific instruction from your teacher.**
- **Always clean up broken glass by using tongs and/or a brush and dustpan.** Discard the pieces in an appropriately labeled "sharps" container.

Safety With Gases

- **Never directly inhale any gas or vapor.** Do not put your nose close to any substance having an odor.
- **Handle materials prone to emit vapors or gases in a well ventilated area.** This work should be done in an approved chemical fume hood.

Lab Rules

- **Never work alone in the laboratory.**
- **Never perform any experiment not specifically assigned by your teacher.** Never work with any unauthorized material.
- **Never consume food or drink, nor apply cosmetics, in the laboratory.** Never store food in the laboratory. Keep hands away from faces. Wash your hands at the conclusion of each laboratory investigation and whenever leaving the laboratory. Remember that some hair products are highly flammable, even after application.
- **NEVER taste chemicals. NEVER touch chemicals.** Even common substances should be considered dangerous, since they can easily be contaminated in the lab.
- **No contact lenses in the lab.** Chemical vapors can get between the lenses and the eyes and cause permanent eye damage.
- **Know the location of all safety/emergency equipment used in the laboratory.** Examples include eyewash stations, safety blankets, safety shower, fire extinguisher, first aid kit, chemical spill kit(s).
- **Know fire drill procedures and the locations of exits.**
- **Know the location of the closest telephone** and be sure there is a posted list of emergency phone numbers, including poison control center, fire department, police, and ambulance.
- **Familiarize yourself with the investigation—especially safety issues—before entering the lab.** Know the potential hazards of the materials, equipment, and the procedures required. Before you start, ask the teacher to explain any parts you do not understand.
- **Before beginning work: tie back long hair, roll up loose sleeves, and put on any required personal protective equipment as required by your teacher.** Avoid or confine loose clothing that could knock things over, ignite from flame, or soak up chemical solutions. Don't wear open-toed shoes or sandals to the lab. If there is a spill, your feet could be injured.
- **Report any accident, incidents, or hazard—no matter how trivial—to your teacher immediately.** Any incident involving bleeding, burns, fainting, chemical exposure, or ingestion should also be reported to your school nurse and/or school physician.
- **In case of fire, alert the teacher and leave the laboratory.**
- **Keep your work area neat and uncluttered.** Only bring books and other materials that are needed to conduct the experiment. Stay at your work area as much as possible. The less movement in a lab, the fewer spills and other accidents will occur.
- **Clean your work area at the conclusion of the experiment, as your teacher directs.**
- **Wash your hands with soap after each investigation.**

Safety Equipment

Have You Got What It Takes?

- **Chemical goggles:** [Meeting ANSI (American National Standards Institute) Standard Z87.1] These should be worn when working with any chemical or chemical solution other than water, when heating substances, when using any mechanical device, or when observing physical processes that could eject an object.

    The wearing of contact lenses for cosmetic reasons should be prohibited in the laboratory. If a student must wear contact lenses prescribed by a physician, that student should wear eye-cup safety goggles meeting ANSI Standard Z87.1 (similar to swimmer's cup goggles).

- **Face shield:** [Meeting ANSI Standard Z87.1] Use in combination with eye goggles when working with corrosives.

- **Eyewash station:** The device must be capable of delivering a copious, gentle flow of water to both eyes for at least 15 minutes. Portable liquid supply devices are not satisfactory and should not be used. A plumbed-in fixture or a perforated spray head on the end of a hose attached to a plumbed-in outlet and designed for use as an eyewash fountain is suitable if it meets ANSI Standard Z358.1 and is within a 30-second walking distance from any spot in the room.

- **Safety Shower:** [Meeting ANSI Standard Z358.1] Location should be within a 30-second walking distance from any spot in the room. Students should be instructed in the use of the safety shower, in the event of a fire or chemical splash on their body that cannot be simply washed off.

- **Gloves:** Polyethylene, neoprene rubber, or disposable plastic may be used. Nitrile or butyl rubber gloves are recommended when handling corrosives.

- **Apron:** Gray or black rubber-coated cloth or vinyl (nylon-coated) halter is recommended.

Prudent Precautions

An Ounce (28.35 g) of Prevention

What would you do if a student dropped a liter bottle of concentrated sulfuric acid? RIGHT NOW? Are you prepared? Could you have altered your handling and storage methods to prevent or lessen the severity of this incident? PLAN now how to effectively react BEFORE you need to. Some planning tips include:

1. Post the phone numbers of your regional Poison Control Center, Fire Department, Police, Ambulance, and Hospital ON your telephone.

2. Practice fire and evacuation drills during labs, and at all times during the year, not just in the fall. Post an evacuation diagram and established evacuation procedure by every entrance to the laboratory.

3. Also have drills on what students MUST do if they are on fire, or have chemical contact or exposure.

4. Mark locations of eyewash stations, safety shower, fire extinguishers (A-B-C tri-class), chemical spill kit, first aid kit, and fire blanket in the laboratory and storeroom. Check for the presence and good working order of this safety equipment prior to conducting each investigation.

5. Lock your laboratory (and storeroom) when you are not present.

6. Compile a MSDS file for all chemicals. This reference resource should be readily accessible in case of a spill or other incident. (Information about Material Safety Data Sheets is on page T27.)

7. Provide for spill control procedures. Handle only those incidents that you FEEL COMFORTABLE in handling. Situations of greater severity should be handled by trained hazardous material responders. In case of such a situation, first contact your EPA administrator for an EPA identification number.

8. Under no circumstances should students fight fires or handle spills.

9. Be sure to recognize and heed the signal words used in most safety labels for materials, equipments, and procedures:
    CAUTION—low level of risk associated with use or in misuse
    WARNING—moderate level of risk associated with use or in misuse
    DANGER—high level of risk associated with use or in misuse

10. Be trained in first aid and basic life support (CPR) procedures. Have first aid kits and spill kits readily available.

11. Before the class begins an investigation, review specific safety rules and demonstrate proper procedures.

12. Never permit students to work in your laboratory without your supervision. No unauthorized investigations should ever be conducted, nor unauthorized materials be brought into the laboratory.

13. Fully document ANY INCIDENT that occurs. Documentation will provide the best defense in terms of liability and it is a critical tool in helping to identify area(s) of laboratory safety that need improvement. Remind students that any safety incident, no matter how trivial, must be reported directly to you.

Safety With Microbes

What You Can't See CAN Hurt You

Pathogenic (disease-causing) microorganisms are not appropriate investigational tools in the high-school laboratory and should never be used.

Consult with the school nurse to screen students who may be receiving immuno-suppressive drug therapy that could lower immune response. Such individuals are extraordinarily sensitive to potential infection from generally harmless microorganisms and should not participate in laboratory investigations unless permitted to do so by a physician. Do not allow students with any open cuts, abrasions, or open sores to work with microorganisms.

Aseptic Technique

Demonstrate correct aseptic technique to students PRIOR to conducting an investigation. Never transfer liquid media by mouth. Wherever possible, use sterile cotton applicator sticks in place of inoculating loops and bunsen burner flames for culture inoculation. Remember to use appropriate precautions when disposing of cotton applicator sticks: they should be autoclaved or sterilized before disposal.

Treat ALL microbes as pathogenic. Seal with tape all petri dishes containing bacterial cultures. Do not use blood agar plates, and never attempt to cultivate microbes from a human or animal source.

Never dispose of microbe cultures without first sterilizing them. Autoclave or steam-sterilize all used cultures and any materials that have come in contact with them at 120°C and 15 psi for 15–20 minutes. If these devices are not available, flood or immerse these articles with or in Clorox® bleach (full-strength) for 30 minutes, then discard. Use the steam sterilizer or autoclave them yourself—do not allow students to use them.

Wash all lab surfaces with a disinfectant solution before and after handling bacterial cultures.

Handling Bacteriological Spills

Never allow students to clean up bacteriological spills. Keep on hand a spill kit containing 500 mL Clorox® bleach (full-strength), biohazard bags (autoclavable), forceps, and paper towels.

In the event of a bacterial spill, cover the area with a layer of paper towels. Wet the paper towels with the bleach solution; allow to stand for 15–20 minutes. Wearing gloves and using forceps, place the residue in the biohazard bag. If broken glass is present, use a brush and dustpan to collect material, and place it in a suitably marked container.

Animal Care

Ethics in the Laboratory

It is recommended that teachers follow the Guidelines for the Use of Live Animals established by the National Association of Biology Teachers. The master disposal/release list provides specific guidelines for taking care of animals after these lab investigations.

NABT GUIDELINES FOR THE USE OF LIVE ANIMALS
(Revised April 1991)

Living things are the subject of biology, and their direct study is an appropriate and necessary part of biology teaching. Textbook instruction alone cannot provide students with a basic understanding of life and life processes. The National Association of Biology Teachers recognizes the importance of research in understanding life processes and providing information on health, disease, medical care, and agriculture.

The abuse of any living organism for experimentation or any other purpose is intolerable in any segment of society. Because biology deals specifically with living things, professional biology educators must be especially cognizant of their responsibility to prevent the inhumane treatment of living organisms in the name of science and research. This responsibility should extend beyond the confines of the teacher's classroom to the rest of the school and community.

The National Association of Biology Teachers believes that students learn the value of living things, and the values of science, by the events they witness in the classroom. Such teaching activities should develop in students and teachers a sense of respect and pleasure in studying the wonders of living things. NABT is committed to providing sound biological education and promoting humane attitudes toward animals. These guidelines should be followed when live animals are used in the classroom:

A. Biological experimentation should be consistent with a respect for life and all living things. Humane treatment and care of animals should be an integral part of any lesson that includes living animals.

B. Exercises and experiments with living things should be within the capabilities of the students involved. The biology teacher should be guided by the following conditions:

1. The lab activity should not cause the undue loss of a vertebrate's life. Bacteria, fungi, protozoans, and invertebrates should be used in activities that may require use of harmful substances or loss of an organism's life. These activities should be clearly supported by an educational rationale and should not be used when alternatives are available.

2. A student's refusal to participate in an activity (e.g., dissection or experiments involving live animals, particularly vertebrates) should be recognized and accommodated with alternative methods of learning. The teacher should work with the student to develop an alternative for obtaining the required knowledge or experience. The alternative activity should require the student to invest a comparable amount of time and effort.

C. Vertebrate animals can be used as experimental organisms in the following situations:

1. Observations of normal living patterns of wild animals in their natural habitat or in zoological parks, gardens, or aquaria.

2. Observations of normal living functions such as feeding, growth, reproduction, activity cycles, etc.

3. Observations of biological phenomena among and between species such as communication, reproductive and life strategies, behavior, interrelationships of organisms, etc.

D. If live vertebrates are to be kept in the classroom, the teacher should be aware of the following responsibilities:

1. The school, under the biology teacher's leadership, should develop a plan on the procurement and ultimate disposition of animals. Animals should not be captured from or released into the wild without the approval of both a responsible wildlife expert and a public health official. Domestic animals and "classroom pets" should be purchased from licensed animal suppliers. They should be healthy and free of diseases that can be transmitted to humans or to other animals.

2. Animals should be provided with sufficient space for normal behavior and postural requirements. Their environment should be free from undue stress such as noise, overcrowding, and disturbance caused by students.

3. Appropriate care—including nutritious food, fresh water, clean housing, and adequate temperature and lighting for the species—should be provided daily, including weekends, holidays, and long school vacations.

4. Teachers should be aware of any student allergies to animals.

5. Students and teachers should immediately report to the school health nurse all scratches, bites, and other injuries, including allergies or illnesses.

6. There should always be supervised care by a teacher competent in caring for animals.

E. Animal studies should always be carried out under the direct supervision of a biology teacher competent in animal care procedures. It is the responsibility of the teacher to ensure that the student has the necessary comprehension for the study. Students and teachers should comply with the following:

1. **Students should not be allowed to perform surgery on living vertebrate animals.** Hence, procedures requiring the administration of anesthesia and euthanasia should not be done in the classroom.

2. Experimental procedures on vertebrates should not use pathogenic microorganisms, ionizing radiation, carcinogens, drugs or chemicals at toxic levels, drugs known to produce adverse or teratogenic effects, pain-causing drugs, alcohol in any form, electric shock, exercise until exhaustion, or other distressing stimuli. No experimental procedures should be attempted that would subject vertebrate animals to pain or distinct discomfort, or interfere with their health in any way.

3. Behavioral studies should use only positive reinforcement techniques.

4. Egg embryos subjected to experimental manipulation should be destroyed 72 hours before normal hatching time.

5. Exceptional original research in the biological or medical sciences involving live vertebrate animals should be carried out under the direct supervision of an animal scientist, e.g., an animal physiologist, or a veterinary or medical researcher, in an appropriate research facility. The research plan should be developed and approved by the animal scientist and reviewed by a humane society professional staff person prior to the start of the research. All professional standards of conduct should be applied, as well as humane care and treatment, and concern for the safety of the animals involved in the project.

6. Students should not be allowed to take animals home to carry out experimental studies.

F. Science fair projects and displays should comply with the following:

1. The use of live animals in science fair projects shall be in accordance with the above guidelines. In addition, no live vertebrate animals shall be used in displays for science fair exhibitions.

2. No animal or animal products from recognized endangered species should be kept and displayed.

Preserved Specimens

Fixatives Don't Distinguish Between Living and Dead Tissue

The following practices are recommended when handling or dissecting any preserved specimen:

1. NEVER dissect road kills or nonpreserved slaughterhouse material. Doing so increases the risk of infection.

2. Wear protective gloves and splash-proof safety goggles at all times when handling preserving fluids, preserved specimens, and during dissection.

3. Wear lab aprons. Use of an old shirt or smock is recommended.

4. Conduct dissection activities in a well-ventilated area.

5. Do not allow preservation or body cavity fluids to come into contact with skin. Fixatives do not distinguish between living or dead tissue.

WARD'S Natural Science Establishment provides specimens that are freeze-dried, and rehydrated in a 10% isopropyl alcohol solution. In these specimens, no other hazardous chemical is present.

Reduction of Free Formaldehyde

Currently, federal regulations mandate a permissible exposure level of 0.75 ppm for formaldehyde. Contact your supplier for an MSDS that details the amount of formaldehyde present, as well as gas-emitting characteristics for individual specimens.

Pre-washing specimens (in a loosely covered container) in running tap water for 1–4 hours will dilute the fixative. Formaldehyde may also be chemically bound (thereby greatly reducing dangers) by immersing washed specimens in a 0.5–1.0% potassium bisulfate solution overnight or by placing them in holding solutions containing 1% phenoxyethanol.

Reagents and Storage

General Guidelines

- Store bulk quantities of chemicals in a safe and secure storeroom, not in the teaching laboratory. Store them in well-ventilated, dry areas protected from sunlight and localized heat. Store by similar hazard characteristics, not alphabetically. (See "Chemical Hazard Classes" and "Chemical Storage" for additional recommendations.)

- Label student reagent containers with the substance's name, and hazard class(es). Be sure to use labeling materials that won't be affected by the reagent or others that will be stored nearby. (See "Chemical Hazard Classes" for additional recommendations.)

- Dispose of hazardous waste chemicals according to federal, state, and local regulations. Refer to the Material Safety Data Sheets (available through your supplier) for recommended disposal procedures. Some disposal information is also included in the Master Disposal List that follows this section. NEVER assume that a reagent can be safely poured down the drain unless you are certain.
- Have a chemical spill kit immediately available. Know the procedures for handling a spill of any chemical used during an investigation or in preparing reagents. Never allow students to clean up hazardous chemical spills.
- Remove all sources of flames, sparks, and heat from the laboratory when any flammable material is being used.

Chemical Record-Keeping

Maintaining a current inventory of chemicals can help you keep track of purchase dates and amounts. In addition, if you use your inventory to help you buy only what you need for the school year, it can help you cut down on storage problems.

The purpose of a Material Safety Data Sheet (MSDS) is to provide readily accessible information on chemical substances commonly used in the science laboratory, or in industry. MSDSs are available from suppliers of chemicals.

MSDSs should be kept on file and referred to BEFORE handling ANY chemical. The MSDSs can also be used to instruct students on chemical hazards, evaluate spill and disposal procedures, and warn of incompatibilities with other chemicals or mixtures.

Each MSDS is divided into the following sections:

I   Material Identification: includes name, common synonyms, reference codes, and precautionary labeling

II Ingredients and Hazards

III Physical Data: includes information such as melting point, boiling point, appearance, odor, density, etc.

IV Fire and Explosion Hazard Data: includes flash point, description of fire-extinguishing media and procedures, and information on unusual fire and explosion hazards

V   Health Hazard Data

VI Reactivity Data

VII Spill, Leak, and Disposal Procedures

VIII Special Protection Information: describes equipment needed for safe use

IX Special Precautions and Comments: describes storage requirements and other notes

WARD'S has a pocket guide (WARD'S Catalog No. 32 M 0002) available that explains in greater detail how to use the Material Safety Data Sheets.

Hazard Classes

The hazards presented by any chemical can be grouped into the following categories:

- **Flammables**
- **Corrosives**
- **Poisons (toxics)**
- **Reactives**

(It is important to keep in mind that a particular chemical may have more than one of these hazards.)

A fifth category, those chemicals that do not possess the above properties, are termed "low hazard" materials. Water is an example of a "low hazard" material. Although these materials may not ordinarily represent a hazard, their presence in the lab requires that they be treated differently than they would be in a kitchen or backyard.

Several precautions should be used with all hazard classes:

- All require the use of goggles, gloves, aprons.
- Minimize the amounts available in the lab (100 mL or less).
- Become familiar with first aid measures for each chemical used.
- Become familiar with incompatibility issues for each chemical used.
- Lab and storeroom cleanliness and personal hygiene are essential when dealing with hazardous materials.
- Keep hazardous chemicals in approved containers that are kept closed and stored away from sunlight and rapid temperature changes.
- Chemicals of each hazard class should be stored away from those in other hazard classes.

Storage

Never store chemicals alphabetically, as this greatly increases the risk of promoting a violent reaction. Take these additional precautions to prevent your chemicals from causing a tragedy.

1. Always lock the storeroom and all its cabinets when not in use.
2. Students should not be allowed in storeroom and preparation areas.
3. Avoid storing chemicals on the floor of the storeroom.
4. Do not store chemicals above eye level or on the top shelf in the storeroom.
5. Be sure shelf assemblies are firmly secured to walls.
6. Provide for anti-roll lips on all shelves.
7. Shelving should be constructed of wood. Metal cabinets and shelves are easily corroded.
8. Avoid metal adjustable shelf supports and clips. They can corrode, causing shelves to collapse.
9. Acids, flammables, poisons, and each classification of oxidizer should be stored in their own locking storage cabinets.
10. Storage should protect chemicals from heat, sunlight, moisture, and physical shock.

Additional Resources

You Can Never Be _Too_ Safe:

Your school district may have more information on safety issues. Some districts have a safety officer responsible for safety throughout their schools. Other possible sources for information include your state education agency, teachers' associations and science teachers' associations, and local colleges or universities. In addition, OSHA has 10 regional and 93 area offices dedicated to monitoring safety issues in the workplace, including those relating to biological and chemical hazards.

American Chemical Society Health and Safety Service

This service will refer inquiries to appropriate resources for finding answers to questions about health and safety.

> American Chemical Society (ACS)
> 1155 Sixteenth Street, N.W.
> Washington, D.C. 20036
> (202) 872-4511

Hazardous Materials Information Exchange (HMIX)

Sponsored by the Federal Emergency Management Agency and the U.S. Department of Transportation, HMIX serves as a reliable on-line database. It can be accessed through an electronic bulletin board, and it provides information regarding instructional material and literature listings, hazardous materials, emergency procedures, and applicable laws and regulations.

HMIX can be accessed by a personal computer having a modem (300, 1200, or 2400 baud) with communication parameters set to no parity, 8 data bits, and 1 bit stop. Dial (312) 972-3275. The bulletin board is available 24 hours a day, seven days a week. The service is available free of charge. You pay only for the telephone call.

Safety Reference Works

Gessner, G.H., ed. Hawley's Condensed Chemical Dictionary. 11th Ed., Van Nostrand Reinhold, 1987. (Revised by N. Irving Sax.)

A Guide to Information Sources Related to the Safety and Management of Laboratory Wastes from Secondary Schools. New York State Environmental Facilities Corp., 1985.

Lefevre, M.J. The First Aid Manual for Chemical Accidents. Dowdwen, 1989. (Revised by Shirley A. Conibeau.)

Pipitone, D., ed. Safe Storage of Laboratory Chemicals. John Wiley, 1984.

Prudent Practices for Disposal of Chemicals from Laboratories. Committee on Hazardous Substances in the Laboratory, National Research Council. National Academy Press, 1983.

Prudent Practices for Handling Hazardous Chemicals in Laboratories. Committee on Hazardous Substances in the Laboratory, National Research Council. National Academy Press, 1981.

Strauss, H. and M. Kaufman, ed. Handbook for Chemical Technicians. McGraw-Hill, 1981.

Ward's MSDS User's Guide. Ward's, 1989.

Master Disposal/Release List

Don't Assume You Can Just Throw It Away

Whenever appropriate, the re-use of materials and life forms is to be encouraged. If students will need the same chemicals next year, keeping excess supplies until then makes good environmental and economical sense.

Please note that the guidelines given here cannot account for all possible situations and regulations that may apply in different places and times. Contact WARD'S at 1-800-962-2660 for more information where indicated.

Life Forms

Aquatic Plants Certain aquatic plants (*Elodea*, among others) should not be released or introduced into local habitats. *Elodea* is regulated as a pest organism in Canada and in parts of the northern U.S.A.

Insects Certain insects (cockroaches and termites, for example) require permits for possession or release. Check with your local office of the Animal and Plant Health Inspection Service (APHIS), the U.S. Department of Agriculture, or contact WARD'S.

Microbes Bacteria, fungi, yeasts, growth media, and any materials that have come into contact with these organisms should not be discarded without prior decontamination (sterilization). A permit may be required (in certain states) in order to acquire certain bacteria (*Agrobacterium* and others). Check with WARD'S.

Microinvertebrates and Protists These organisms may be freely released in aquatic environments. Do not release nematodes unless specifically directed to do so.

Macroinvertebrates and Vertebrates Exotic or nonindigenous forms (such as tarantulas, anoles, crayfish, goldfish, or carp) should not be released into local environments. These organisms may not survive climatic conditions or may interfere with native fauna/flora. Contact your local APHIS office or WARD'S for specific information about whether a particular organism would be considered nonindigenous to your area. In some cases, you may be able to find a home for an animal at or through the local pet shop.

Plants Locally cultivated native plants may be introduced by replanting. Ornamental plants that are not cultivated locally should not be introduced into native habitat. Some states (especially California) have strict rules to prevent introduction of nonindigenous plants. Usually plants shipped into these states must pass inspection.

Chemical and Reagent Disposal

WARD'S materials are pre-measured, and provided in low concentrations, to minimize hazards and disposal issues.

Before preparing or disposing of any of the material listed below, familiarize yourself with safety and handling procedures and storage information listed in **Reagents and Storage** in the safety section.

You should be aware of local, state, and federal regulations governing disposal of hazardous materials. **The disposal procedures outlined below should only be used if they conform to these regulations.** Contact a licensed treatment, storage, and disposal facility for disposal of bulk or large quantities of hazardous chemicals. Disposal protocols outlined below are ONLY for the substances (and quantities) specified.

Unless your school's drains are connected to a sanitary sewer system and a treatment plant, no chemicals should ever go down the drain. **NEVER pour chemicals down the drain if your drains empty into a septic system or storm sewer.** Even if you are connected to a sanitary sewer, do not wash any chemical down the drain until you are certain it is safe.

The chemicals in this list are described by a color-code for their hazard class. Remember to store chemicals of different hazard classes away from each other. YELLOW: Reactive RED: Flammable WHITE: Corrosive BLUE: Toxic GREEN: Low Hazard

DISPOSAL: METHOD A—INERT SOLID WASTES (Low Hazard)

Items using DISPOSAL: METHOD A can be considered to be LOW HAZARD materials as long as they remain within the laboratory. Avoid creating or breathing dusts of these materials and any chemical. Their storage code is GREEN, for general storage. Check local, state, and federal regulations to be certain that they can be disposed of in local landfills.

DISPOSAL: METHOD B—SMALL QUANTITIES OF LIQUID WASTES (Low Hazard)

Items using DISPOSAL: METHOD B can be considered to be LOW HAZARD materials as long as they remain within the laboratory. Their storage code is GREEN, for general storage. Wear PPE (personal protective equipment): goggles, apron, and nitrile gloves when working with these materials. Also be sure an eyewash station is nearby. In volumes less than 250 mL, these substances can be disposed of by diluting them 1:20 with water. Place a beaker of the diluted solution in the sink and run water to overflowing for 10 minutes, flushing to sanitary sewer.

Acetylcholine bromide solution (1:10,000) DISPOSAL: METHOD B

Adrenalin chloride solutions (0.001%, 0.0001%, 0.00001%) DISPOSAL: METHOD B

Albumin DISPOSAL: METHOD A

Algae culture solution (Bristol's) DISPOSAL: METHOD B

Alpha amylase DISPOSAL: METHOD A

Ascorbic acid Protect from light. DISPOSAL: METHOD A

ATP powder (Adenosine 5´-triphosphate, disodium salt) Store below 0°C. DISPOSAL: METHOD A

Baking powder (sodium bicarbonate, tartaric acid, and cornstarch) DISPOSAL: METHOD A

Benedict's solution DISPOSAL: METHOD B

Biuret reagent DISPOSAL: METHOD B

Bromothymol blue stain solution (0.1%) DISPOSAL: METHOD B

Cobalt chloride paper (cobaltous chloride, hexahydrate)

> STORAGE: GREEN (general storage). Avoid contact with moisture. **Do not ingest— may affect health.**

> DISPOSAL: Dispose of as inert solid waste. (Quantities of cobalt are so small they can be considered inconsequential for disposal issues.)

Dechlorinator (sodium thiosulfate pentahydrate [hypo] solution, 0.1 M) DISPOSAL: METHOD B

Detain™ (nontoxic) DISPOSAL: METHOD A

Detergent (powder with/without phosphates) DISPOSAL: METHOD B

Ethyl alcohol—denatured: anhydrous or 95%

> STORAGE: RED (flammable liquid)

> DISPOSAL: Wear PPE: goggles, apron, and nitrile gloves. Dilute small volumes (<250 mL) 1:20 with water. Place beaker of diluted solution in the sink and run water to overflowing for 10 minutes, flushing to sanitary sewer.

Euglena medium DISPOSAL: METHOD B

Extraction of Bacterial DNA activity

See individual chemical components: Ethyl alcohol (denatured), Schiff's reagent, Macromolecular solution (inert solid waste)

Fertilizer (commercial; powder)

> STORAGE: GREEN (general storage). Over-pack in sealed plastic bag or plastic container.

> DISPOSAL: Dilute according to instructions on the package and apply to lawn.

Firefly lantern extract

> WARNING: ALWAYS wear PPE: disposable polyethylene gloves, apron, and goggles. This material is a POISON—may be harmful by ingestion or skin absorption. Prolonged or repeated exposure may cause allergic reactions in certain sensitive individuals. **READ MSDS and product label before use.**

> STORAGE: BLUE (poison). Store below 0°C. Powder consists of 47% firefly lanterns, 42% potassium dihydrogen arsenate, 11% magnesium sulfate.

> DISPOSAL: Dispose of in an approved chemical landfill through a certified facility for chemical waste treatment, storage, and disposal. Facilities must be certified through the Resource Conservation and Recovery Act.

Gibberellic acid solution (1:1,000,000)

> PREPARATION: Wear PPE: goggles, apron, and nitrile gloves. Avoid creation of and breathing of dusts. Mix 0.0001 g of gibberellic acid to 100 mL of distilled water. Mix well.

STORAGE: GREEN (general storage). Very deliquescent. Keep container tightly closed.

DISPOSAL: Wear PPE: goggles, apron, and nitrile gloves. Mix up to 1g in 1L of distilled water. Place beaker of diluted solution in the sink and run water to overflowing for 10 minutes, flushing to sanitary sewer.

Hydrochloric acid—dilute

PREPARATION: Purchase ready-made or dilute from concentrated HCl. To dilute wear PPE: face shield, splashproof safety goggles, nitrile gloves, and rubber apron. Work inside an approved chemical fume hood. Be sure an eyewash station and safety shower are within a 15-second walk from dispensing activities. Slowly add 8 mL of concentrated HCl to 500 mL of distilled water in a volumetric flask and fill to the 1000-mL mark.

STORAGE: GREEN (general storage) for dilute; WHITE (corrosive liquid) for concentrated

DISPOSAL: Wear PPE: goggles, apron, and nitrile gloves. Neutralize only small amounts (<250 mL) at any one time. Slowly add 1 M sodium bicarbonate to the acid until neutrality is confirmed by litmus test. Place beaker of neutralized solution in sink and run water to overflowing for 10 minutes, flushing to a sanitary sewer.

3-Indoleacetic acid solutions (1:10,000; 1:1,000,000)

PREPARATION: Wear PPE: goggles, apron, and nitrile gloves. Avoid creating and breathing dusts. 1:10,000 solution—mix 0.01 mL of distilled IAA with 50 mL of ethyl alcohol; stir until dissolved. Add distilled water, diluting to the 100-mL mark. Mix well.

1:1,000,000 solution—mix 0.001 mL of distilled IAA with 50 mL of ethyl alcohol; stir until dissolved. Add distilled water, diluting to the 100-mL mark. Mix well.

STORAGE: GREEN (general storage)

DISPOSAL: Dilute 1:20 with water. Place beaker of diluted solution in sink and run water to overflowing for 10 minutes, flushing to sanitary sewer.

Indophenol solution (0.1%) DISPOSAL: METHOD B

Lactose-intolerance milk treatment (liquid lactase solution) DISPOSAL: METHOD A

Limewater (calcium hydroxide solution—0.14%)

PREPARATION: Wear PPE: goggles, apron, and nitrile gloves. Avoid dusting conditions. Carefully dissolve 0.14 g of calcium hydroxide in 100 mL of distilled water (saturation limit at 25°C).

STORAGE: GREEN (general storage)

DISPOSAL: Place in beaker in the sink and run water to overflowing for 10 minutes, flushing to sanitary sewer.

Lugol's iodine solution

STORAGE: BLUE (poison)

SPILL: Wear PPE: goggles, apron, and nitrile gloves. Absorb onto paper towels. Wipe spill with 0.1 M sodium thiosulfate solution. (You may be using this solution as a dechlorinator)

DISPOSAL: Wear PPE: goggles, apron, and nitrile gloves. Place 250 mL in a 1-L beaker. Slowly add 0.1 M sodium thiosulfate and mix until the solution is decolorized (fully reduced). Place beaker of decolorized solution in the sink and run water to overflowing for 10 minutes, flushing to sanitary sewer.

Macromolecular solution—WARD'S DISPOSAL: METHOD A

Malt agar (dehydrated) DISPOSAL: METHOD A

Nail polish (clear) STORAGE: RED (flammable liquid)

Nicotine solution DISPOSAL: METHOD B

Nutrient agar (plates)

STORAGE: GREEN (general storage). Store under refrigeration (6°C) until needed.

DISPOSAL: LOW HAZARD for laboratory handling when sterile (autoclave at 121°C; 15 psi for 15–20 minutes). May also be chemically sterilized by applying a thin layer of household bleach (5% sodium hypochlorite) to plate surface, and waiting 20 minutes. Always tape plates shut before discarding as inert solid waste.

Nutrient broth (prepared; sterile)

STORAGE: GREEN (general storage). Store under refrigeration (6°C) until needed.

DISPOSAL: LOW HAZARD for laboratory handling when sterile. Autoclave or steam-sterilize (121°C; 15 psi for 15–20 minutes) contaminated material.

Oil, refined (virgin)

STORAGE: RED (flammable liquid)

DISPOSAL: According to state/local regulations, usually at service stations accepting waste oil.

Pancreatin solution (10%) DISPOSAL: METHOD B

Pepsin solution (10%) DISPOSAL: METHOD B

Phenol red (sodium salt)

STORAGE: GREEN (general storage)

HANDLING: Avoid creating and breathing dusts. Wear particle mask when making solutions for laboratory handling.

Plant mineral solutions

Individual chemical components: calcium chloride (1.0 M); magnesium chloride (1.0 M); iron chelate (0.5 M); potassium chloride (1.0 M); potassium phosphate (1.0 M); sodium phosphate (1.0 M); sodium sulfate (1.0 M); trace element solution; sodium nitrate (1.0 M); calcium nitrate (1.0 M); magnesium sulfate (1.0 M); potassium nitrate (1.0 M).
DISPOSAL: METHOD B

Quinine sulfate solution (quinine bisulfate solution)—0.1% DISPOSAL: METHOD B

Schiff's reagent

STORAGE: GREEN. Store either frozen or at refrigeration temperatures.

DISPOSAL: Wear PPE: goggles and apron. Dissolve unused contents of vial 1:20 in water. Place beaker of diluted solution in the sink and run water to overflowing, flushing to sanitary sewer.

Simulated blood—WARD'S May be diluted for discharge to sanitary sewer.
DISPOSAL: METHOD A

Sodium bicarbonate solution (1.0 M)—for neutralizing acids

STORAGE: GREEN (general storage)

PREPARATION: Wear PPE: goggles and apron. Dissolve 84 g in 500 mL distilled water in a volumetric flask and fill to the 1-L mark.

USE AND DISPOSAL: Add slowly to acidic solutions until neutrality is confirmed by litmus test. *Never neutralize an acid solution more concentrated than 1M.* Place beaker of neutralized solution in the sink and run water to overflowing for 10 minutes, flushing to sanitary sewer.

Sodium bisulfate solution (0.5 M) not for student use **— for neutralizing sodium hydroxide solution (0.1 M)**

PREPARATION: Wear PPE: goggles, face shield, and nitrile gloves. Avoid creating or breathing dusts. Dissolve 69 g of sodium bisulfate in 500 mL of distilled water in a volumetric flask and fill to 1000-mL mark.

STORAGE: WHITE (corrosive liquid). Aqueous solutions are strongly acidic.

USE AND DISPOSAL: Place 250 mL in a 1-L beaker and slowly add 0.1 M of NaOH until neutrality is confirmed by litmus test. Place beaker of neutralized solution in the sink and run water to overflowing for 10 minutes, flushing to sanitary sewer.

Sodium hydroxide solution (0.1 M)

PREPARATION: Wear PPE: goggles, face shield, nitrile gloves, and a corrosive-resistant apron. Work only under an approved chemical fume hood. Be sure an eyewash station is in close proximity to dispensing activities. Place 4 g of NaOH pellets in a 1-L flask and slowly add 500 mL of distilled water with constant swirling until dissolved. Avoid creating and breathing dusts. Add enough distilled water to bring total volume to 1000 mL. Label: *CAUTION: Irritant.*

STORAGE: GREEN (general storage) for solution; WHITE (corrosive solid) for pellets. Avoid contact with moisture, acids, acid fumes, metals, alloys and organics.

DISPOSAL: Wear PPE: goggles, apron, nitrile gloves. Place 250 mL in a 1-L beaker. Slowly add 0.5 M of sodium bisulfate until neutrality is confirmed by litmus test. Place beaker of neutrlized solution in the sink and run water to overflowing for 10 minutes, flushing to sanitary sewer.

Sodium thiosulfate pentahydrate (hypo)

STORAGE: GREEN (general storage). Avoid contact with strong oxidizers, acids, or water-reactive materials.

Sodium thiosulfate (hypo) solution (0.1 M) — for neutralizing Lugol's or as Dechlorinator

PREPARATION: Wear PPE: goggles, apron, and nitrile gloves. Avoid creating and breathing dusts. Dissolve 25 g of solid sodium thiosulfate in 500 mL of distilled water in a 1-L volumetric flask. Then add distilled water to 1000-mL mark. Label: CAUTION: Irritant.

STORAGE: GREEN (general storage)

DISPOSAL: Wear PPE: goggles, apron, and nitrile gloves. Place beaker of solution in the sink and run water to overflowing for 10 minutes, flushing to sanitary sewer.

Starch, soluble DISPOSAL: METHOD A

Sucrose DISPOSAL: METHOD A

Trace element solution CONTENTS: Manganese chloride (0.001%); boric acid (0.002%); zinc chloride (0.0001%); cupric chloride (<0.001%) ; sodium molybdate (<0.001%); water (<99.9%) DISPOSAL: METHOD B

Master Materials List

This list indicates all supplies needed to perform the lab investigations for a class of 30 working in pairs.

WARD'S is the exclusive supplier for the *Biology: Visualizing Life* program. Catalog numbers for WARD'S are in parentheses following each item. The quantities to order are provided in "Materials List, by Investigation."

WARD'S Natural Science Establishment, Inc.
5100 W. Henrietta
P.O. Box 92912
Rochester, NY 14692-9012

Phone: 1-800-962-2660
Fax: 1-800-635-8439

Investigation Kits

WARD's has created kits for 32 students containing the expendable supplies for each lab manual investigation, shown below. Equipment and supplies may also be ordered separately, using the charts that follow.

Investigation	WARD'S No.	Investigation	WARD'S No.
1–1 The Compound Microscope	(36 M 0201)	16–2 Bactericidal Effect of Soap	(36 M 0222)
1–2 Crime Lab	(36 M 0202)	17–1 Observing Pond Water	(36 M 0223)
2–1 Designing Control Experiments	(36 M 0203)	17–2 Prokaryotic and Eukaryotic Algae	(36 M 0224)
2–2 Diversity of Life	(36 M 0204)	18–1 Plant Growth and Soil Conditions	(36 M 0225)
3–1 Cytoplasm and Organelles	(36 M 0205)	19–1 Stomata and Transpiration Rates	(36 M 0226)
3–2 Structure and Function of Cells	(36 M 0206)	20–1 Vitamin C From Foods	(36 M 0227)
4–1 Diffusion and Cell Membranes	(36 M 0207)	20–2 Regulatory Chemicals of Plants	(36 M 0228)
5–1 Plant and Animal Interrelationships	(36 M 0208)	21–1 Roundworms and Earthworms	(36 M 0229)
5–2 Release of Energy	(36 M 0209)	22–1 Insect Behavior and Bioluminescence	(36 M 0230)
6–1 The Case of the Long-Lost Son	(36 M 0210)	23–1 Snails	(36 M 0231)
6–2 A Human Pedigree	(36 M 0211)	24–1 Live Crickets	(36 M 0232)
7–1 Protein Synthesis Drama	(36 M 0212)	24–2 Comparison of Five Arthropod Classes	(36 M 0233)
8–1 Investigating Human Karyotypes	(36 M 0213)	25–1 Fish Morphology and Behavior	(36 M 0234)
8–2 Transforming Genetic Information	(36 M 0214)	26–1 Field Identification of Flying Birds	(36 M 0235)
9–1 Peppered Moth Survey	(36 M 0215)	27–1 Sweat Gland Activity	(36 M 0236)
10–1 Animals of the Future	(N/A)	28–1 Taste and Smell Sense Perception	(36 M 0237)
11–1 Human Evolution	(36 M 0216)	29–1 Effects of Hormones on Circulation	(36 M 0238)
12–1 Mapping an Environmental Site	(36 M 0217)	30–1 How Drugs Affect the Heartbeat Rate	(36 M 0239)
13–1 Detergents as Pollutants	(36 M 0218)	31–1 Normal and Sickled Red Blood Cells	(36 M 0240)
14–1 Acid Rain Effects	(36 M 0219)	32–1 Disease Transmission Simulation	(36 M 0241)
14–2 Oil-Degrading Microbes	(36 M 0220)	33–1 Enzyme Action in Digestion	(36 M 0242)
15–1 Classification	(N/A)	33–2 Lactose Digestion	(36 M 0243)
16–1 Antibiotics and Zones of Inhibition	(36 M 0221)	34–1 Embryonic Development	(36 M 0244)

Biological Supplies (includes living organisms)

Item	Inv.	Item	Inv.
Amoeba culture (live), vital stained (87 M 0380)	3–1	Chick Embryo Slide, 33-Hour (92 M 9040)	34–1
Anabaena culture (live) (86 M 1800)	17–2	Chick Embryo Slide, 43-Hour (92 M 9058)	34–1
Aquatic Plants, floating (live), jar (86 M 7650)	19–1	Chick Embryo Slide, 56-Hour (92 M 9070)	34–1
Bacillus subtilis (live), with growth media (85 M 1840)	16–1	Chiton (preserved), pkg. of 10 (68 M 7002)	2–2
		Chlorella culture (live) (86 M 0126)	17–2
Blood Cells (normal), slide (93 M 6540)	31–1	Crayfish (live), pkg. of 12 (87 M 6031)	24–2
Blood Cells (sickle-cell), slide (93 M 8120)	31–1	Crickets (live), pkg. of 50 (87 M 6102)	24–1, 24–2
Blood (simulated), 5 mL (250 M 0011)	1–2	*Daphnia magna* (live) (87 M 5210)	2–2, 29–1, 30–1
Carp (live) (87 M 8103)	25–1	E. coli (live), with growth media (85 M 1860)	16–1
Centipede (preserved), pkg. of 10 (68 M 3112)	24–2	Earthworms (live), pkg. of 10 (87 M 4660)	21–1
Chick Embryo Slide, 28-Hour (92 M 9035)	34–1	Eel, Vinegar, culture (live) (87 M 2900)	21–1

Item	Inv.	Item	Inv.
Elodea (live), pkg. of 10 (86 M 7500)	3–1, 5–1	Sea Urchin Slide (all stages) (92 M 8330)	34–1
Euglena sp. (live) (87 M 0100)	2–2, 8–2	Seeds, Bean (untreated) (86 M 8016)	2–1
Fish Scale Slide Set (95 M 0825)	25–1	Seeds, Lima Bean, pkg. of 50 (86 M 8008)	20–2
Frog Blood, slide (92 M 3640)	3–2	Seeds, Radish (86 M 8280)	14–1
Frog Skin, slide (92 M 3643)	3–2	Seeds, Sunflower (86 M 8320)	18–1
Frog Sperm, slide (92 M 8803)	3–2	Seeds, Tomato (86 M 8340)	18–1
Goldfish (live), pkg. of 12 (87 M 8100)	25–1	Snails, Land (live), pkg. of 6 (87 M 4306)	23–1
Millipedes (live), pkg. of 12 (87 M 6971)	24–2	Snails, Pond (live), pkg. of 12 (87 M 4141)	5–1
Nostoc, ball colonies (live) (86 M 2155)	2–2, 17–2	*Spirogyra* culture (live) (86 M 0658)	13–1
Oil-Degrading Microbes Set (live) (85 M 7000)	14–2	Tapeworm, adult (preserved) (68 M 0551)	2–2
Pillbugs (live), pkg. of 45 (87 M 5520)	24–2	Tarantula (live) (87 M 6965)	24–2
Pond Water, Dry Mix (87 M 9055)	17–1	Tarantula (preserved) (68 M 3071)	24–2
Saccharomyces cerevisiae, culture (live) (85 M 5000)	2–2		

Chemicals and Media

Some items in this category have limited shelf lives.

Item	Inv.	Item	Inv.
Acetylcholine Solution, 100 mL (38 M 2021)	30–1	Indoleacetic Acid, 5 g (39 M 1661)	20–2
Adrenaline Solution, 100 mL (38 M 2053)	29–1	Indophenol Solution, 0.1%, 500 mL (37 M 9542)	20–1
Algae Culture Solution (makes 1 L) (88 M 3250)	13–1	Limewater (calcium hydroxide solution), 4 L (37 M 2631)	5–2
Amylase, 20 g (38 M 0058)	33–1	Litmus Paper (blue), pkg. of 1,200 strips (15 M 3105)	33–1
Antibiotic Discs, control, pkg. of 100 (38 M 1600)	16–1	Litmus Paper (red), pkg. of 1,200 strips (15 M 3107)	33–1
Antibiotic Discs, Erythromycin, pkg. of 50 (38 M 1605)	16–1	Lugol's Iodine Solution (39 M 1685)	17–2, 33–1, 33–2
Antibiotic Discs, Kanamycin, pkg. of 50 (38 M 1606)	16–1	Microbial Growth Measurement Strips, pkg. of 25 (15 M 1996)	14–2
Antibiotic Discs, Tetracycline, pkg. of 50 (38 M 1611)	16–1	Mineral Solutions (266 M 8400)	18–1
Ascorbic Acid, 100 g (39 M 2006)	20–1	Nicotine Solution, 100 mL (39 M 2457)	30–1
ATP Powder, 1 g (39 M 0174)	22–1	Pancreatin, 100 g (30 M 2728)	33–1
Benedict's Solution (37 M 0698)	33–2	Pepsin, 100 g (39 M 2866)	33–1
Bromothymol Blue, 0.1%, 500 mL (38 M 9100)	5–1	pH Indicator Strips, 200 (15 M 2505)	14–1
Cobalt Chloride Test Paper, 12 sheets (37 M 1340)	19–1	Phenol Red, 5 g (38 M 9431)	32–1
Detain™ (37 M 7950)	8–2	Plates, Nutrient Agar, pkg. of 6 (88 M 0905)	16–1, 16–2
Ethyl Alcohol, 95% (39 M 0283)	27–1, 30–1	Quinine Sulfate Solution, 0.1%, 500 mL (37 M 9545)	28–1
Euglena Medium, 1 L (88 M 5200)	8–2	Refined Oil, 4 oz. (750 M 3502)	14–2
Firefly Lantern Extract, 50 mg (37 M 1580)	22–1	Soap, Antibacterial with Iodine, 16 oz. (15 M 9829)	16–2, 27–1, 33–1
Gibberellic Acid, 5 g (39 M 1446)	20–2	Sodium Hydroxide, dilute, 1 L (37 M 9546)	33–1
Glucose, 500 g (39 M 1357)	33–2	Starch, soluble, 100 g (39 M 3275)	17–2, 33–1
Glucose Test Papers, pkg. of 50 strips (14 M 4107)	33–2	Water, distilled (88 M 7005)	14–2, 18–1, 22–1
Hydrochloric Acid, 0.1 M, 1 L (37 M 9561)	14–1, 32–1, 33–1		

Laboratory Equipment

Item	Inv.	Item	Inv.
Aquarium/Terrarium, 1 gal. (21 M 2101)	13–1, 23–1, 24–2	Dropper Bottle, 1 oz. (17 M 1451)	32–1
Balance, Triple Beam(15 M 6057)	4–1, 13–1	Erlenmeyer Flask, 250 mL (17 M 2982)	5–2, 20–2
Beakers, 600 mL (17 M 4060)	4–1, 21–1, 24–2, 24–1, 25–1, 33–1	Filter Papers, 9 cm, pkg. of 100 (15 M 2835)	14–1
Beakers, 250 mL (17 M 4040)	4–1, 33–2	Fish Net (21 M 3702)	25–1
Beaker, 150 mL (17 M 4030)	1–1	Flashlight (pen-size) (15 M 4000)	25–1
Beaker, 100 mL (17 M 4020)	8–2, 30–1	Fluorescent Light (bulbs included) (20 M 5100)	13–1
Beakers, 50 mL (17 M 4010)	14–1	Forceps (14 M 0999)	1–1, 16–1, 25–1
Bottle, Vacuum (insulated) (•)	5–2	Germination Tray (20 M 3230)	18–1, 20–2
Container (15 M 0539)	2–2	Glass Rods, pkg. of 10 (17 M 6010)	28–1
Coverslips, pkg. of 100 (14 M 3555)	throughout	Glass Tubing (17 M 0931)	5–2
Culture Dish (17 M 0560)	17–1, 21–1	Gloves, disposable, box of 100 (15 M 1071)	throughout
Culture Tube and Cap, 16 × 150 mm (17 M 1340)	5–1, 14–2	Graduated Cylinder, 10 mL (17 M 0170)	8–2, 14–1, 33–1
Dechlorinator (21 M 2292)	5–1, 13–1, 24–2, 25–1	Graduated Cylinder, 25 mL (17 M 0171)	28–1
Dissection Pan Set (14 M 7011)	21–1	Hand Lens (25 M 1350)	21–1, 23–1

Item	Inv.
Hot Plate (15 M 8055)	20–1, 33–1, 33–2
Ice Bucket (18 M 1625)	24–1
Incubator (15 M 0060)	8–2, 16–1, 16–2
Light Bulb (150 W, for light source)	
(36 M 4173)	5–1, 21–1, 23–1, 24–1, 27–1
Light Source (36 M 4168)	5–1, 21–1, 23–1, 24–1, 27–1
Meter Stick (15 M 4065)	12–1
Microscope, Compound Light	
(24 M 2310)	throughout
Microscope Slides, pkg. of 72	
(14 M 3500)	throughout
Microscope Slides, concave, pkg. of 12	
(14 M 3510)	21–1, 29–1, 30–1, 33–2
Personal Safety Set	
antifog goggles	
cotton-lined latex gloves	
vinyl apron (15 M 3041)	throughout
Petri Dishes, disposable, pkg. of 20	
(18 M 7101)	2–1, 14–1, 14–2, 23–1, 29–1
Pipets, transfer, pkg. of 500	
(18 M 2971)	29–1

Item	Inv.
Probe (14 M 0958)	24–2
Rubber Tubing (50') (15 M 1138)	5–2
Scalpel (14 M 0705)	19–1, 20–2, 33–1
Scissors (14 M 0988)	1–1, 7–1, 8–1
Spot Plate (15 M 3980)	17–2
Stereomicroscope	
(24 M 4602)	21–1, 23–1, 24–1, 25–1, 34–1
Stoppers, size 00, pkg. of 110	
(15 M 8459)	33–1
Stoppers, 2-hole, size 9, pkg. of 10	
(15 M 8519)	5–2
Swabs, sterile, pkg. of 100	
(14 M 5502)	16–1, 27–1, 28–1
Test Tube, 15 × 125 mm	
(17 M 0620)	20–1, 32–1, 33–1, 33–2
Test Tube Rack (18 M 0010)	14–2, 20–1, 33–1
Thermometer (15 M 1462)	33–1
Watch Glass, Syracuse, pkg. of 12	
(17 M 0530)	17–2, 24–2

Miscellaneous

Substances with a bullet (•) indicate materials not available through WARD'S. These items can be purchased separately at a grocery store or brought from home.

Item	Inv.
Antacid, bottle of 100 (•)	30–1
Apple, fresh red (•)	14–1, 24–1, 28–1
Aspirin, bottle of 100 (•)	30–1
Baking Powder, box (37 M 2270)	1–2
Box (shoe box, etc.) (•)	23–1, 24–1, 25–1
Calculator (27 M 3055)	11–1
Cloth (•)	22–1
Cloth Fibers (•)	1–2
Cloth Square, 35 × 35 cm (15 M 2538)	24–1
Coffee (1 cup) (•)	30–1
Common Items, sets (see Investigation 15-1) (•)	15–1
Construction Paper, 50 sheets (15 M 9841)	21–1
Containers with Lids, 1 qt. (18 M 9919)	18–1
Corks, pkg. of 100 (15 M 8364)	18–1
Corn Meal (•)	1–2
Corn Syrup, 500 mL (39 M 1461)	4–1
Cotton, 1 lb. (15 M 3830)	14–1, 18–1, 20–2
Cough Drops, package (•)	30–1
Detergent Powder (with phosphates)	
(15 M 1287)	1–2, 13–1
Detergent Powder (without phosphates)	
(15 M 1288)	1–2, 13–1
Fabric (dark) (15 M 2539)	22–1
Flour (•)	1–2
Eggs, fresh, 12 (•)	4–1, 33–1
Fur (from a pet animal) (•)	1–2
Glue, white, 4 oz. (15 M 9806)	12–1
Hair (student) (•)	1–2
Index Cards, pkg. of 1,000 (15 M 9807)	7–1
Insects (•)	22–1
Jars, clear plastic with lid (18 M 7192)	22–1
Juice, assorted varieties	
(see Investigation 20-1) (•)	20–1
Labels, pkg. of 240 (15 M 1832)	4–1, 20–2
Lactose-Intolerance Product (liquid)	
(39 M 0140)	33–2
Lens Paper, box of 50 (15 M 8250)	1–1, 3–2, 17–2
Marker Set, 4 colors (15 M 4635)	4–1, 7–1, 12–1, 28–1
Medicine Droppers, pkg. of 12	
(17 M 0230)	throughout
Milk, 1 pt. (•)	33–2
Molasses, 350 mL (39 M 2298)	5–2

Item	Inv.
Nail Polish, clear (37 M 1862)	19–1
Nasal Spray (•)	30–1
Needle, Sewing (•)	22–1
Newspaper (•)	1–1
Onion, fresh (•)	28–1
Paper Clips, pkg. of 1,000 (15 M 9815)	7–1, 19–1
Paper, graph (15 M 3835)	5–2, 9–1
Paper Grocery Bags (•)	7–1
Paper Towels, roll (15 M 9844)	throughout
Paper, white (•)	5–1, 6–2, 6–3, 8–1
Paper, erasable bond (•)	27–1
Pencils, colored, pkg. of 12 (15 M 2576)	9–1, 31–1
Pencils, No. 2, pkg. of 12	
(15 M 9816)	6–2, 6–3, 23–1, 26–1
Plastic Bags, zipper type, pkg. of 10	
(18 M 6924)	19–1, 24–1
Plastic Cups, pkg. of 25 (18 M 3675)	28–1
Plastic Wrap, roll (15 M 9858)	13–1
Poster Board (18 M 2105)	12–1
Pot, 3 in., pkg. of 10 (20 M 2130)	22–1
Potato, fresh (•)	28–1
Potting Soil, 8 lb. bag (20 M 8306)	22–1
Protractor (15 M 4067)	11–1
Ruler, metric	
(14 M 0810)	2–2, 11–1, 16–1, 18–1, 20–2, 27–1, 31–1
Salt (sodium chloride), 500 g (37 M 5480)	1–2, 28–1
Sand, 500 g (20 M 7423)	1–2
Screen, metal window, 1 yd.	
(15 M 0002)	22–1, 24–1, 24–2
Scrub Brush (for hands) (15 M 9106)	16–2
Soap, bar (•)	16–2
Spoon, plastic (•)	28–1, 29–1
Stakes, pkg. of 12 (21 M 0364)	12–1
Steel Wool, 16 pads (15 M 8798)	14–1
Sticker Tag Set, 8 colors (15 M 9805)	12–1
String (15 M 9863)	12–1
Sucrose, 500 g (39 M 3182)	5–2, 28–1
Tape, cellophane, roll	
(15 M 1957)	7–1, 8–1, 16–2, 19–1, 21–1
Tape, masking, roll (15 M 9828)	24–1
Tea, brewed (1 cup) (•)	30–1
Terrestrial Plant (Geranium) (86 M 6900)	19–1

Item	Inv.	Item	Inv.
Thread, spool, pkg. of 3 (15 M 9837)	22–1	Vinegar, (acetic acid solution), 1 pt.	
Toothpicks, pkg. of 750 (15 M 9864)	8–2, 33–2	(39 M 0138)	4–1, 28–1
Trowel (20 M 7015)	20–2	Vinegar, Cider, 1 pt. (39 M 0139)	21–1
Twist Ties, 250' roll (14 M 0947)	19–1	Watch with second hand (15 M 0512)	19–1, 29–1, 30–1
Vegetable Oil, 500 mL (37 M 9539)	33–1	Wax Pencil (15 M 1155)	throughout
Vermiculite, 14 lb. bag (20 M 8630)	18–1, 20–2	Yeast, Baker's (dry), 100 g (38 M 5826)	5–2

Enrichment

Additional related materials available from WARD'S.

Item	Inv.	Item	Inv.
Cell Structure and Function (color slides)		Chromosome Spread: Turner's Syndrome	
(170 M 9109)	3–2	(33 M 1051)	8–1
The Cell (transparency set) (75 M 0110)	3–2	Human Karyotype Form (33 M 1045)	8–1
Landmarks of the Human Genome, Macintosh®		Dichotomous Key to Pond Microlife, Apple II®	
(74 M 3673)	8–1	(74 M 1858)	17–1
Landmarks of the Human Genome, IBM®		Dichotomous Key to Pond Microlife, Macintosh®	
(74 M 3674)	8–1	(74 M 0041)	17–1
Chromosome Spread: Down Syndrome Male		Plant Mineral Deficiency Activity (20 M 8400)	18–1
(33 M 1048)	8–1	Field Guide to Birds, Eastern Region (32 M 2103)	26–1
Chromosome Spread: Down Syndrome Female		Field Guide to Birds, Western Region (32 M 2104)	26–1
(33 M 1049)	8–1		
Chromosome Spread: Klinefelter's Syndrome			
(33 M 1050)	8–1		

Materials List, by Investigation

This list indicates the materials needed for a class of 30 working in pairs to perform the lab manual investigations. Catalog numbers for WARD'S and the quantities to order are provided. In a few cases, some common household items are not available through WARD'S. For such items, no serial number is given. These items can be purchased separately at a grocery store or brought from home.

Investigation 1–1 (pp. 69–76)

Lab Equipment

Beakers, 150 mL (17 M 4030)	15
Coverslips, pkg. of 100 (14 M 3555)	1
Forceps (14 M 0999)	15
Microscope, Compound Light (24 M 2310)	15
Microscope Slides, pkg. of 72 (14 M 3500)	1
Scissors (14 M 0988)	15

Miscellaneous

Lens Paper, box of 50 (15 M 8250)	15
Medicine Droppers, pkg. of 12 (17 M 0230)	2
Newspaper (•)	1

Investigation 1–2 (pp. 77–80)

Biological Supplies

Blood (simulated), 5 mL (250 M 0011)	1

Lab Equipment

Coverslips, pkg. of 100 (14 M 3555)	10
Microscope, Compound Light (24 M 2310)	15
Microscope Slides, pkg. of 72 (14 M 3500)	1

Miscellaneous

Baking Powder, box (37 M 2270)	1
Cloth Fibers (•)	15
Corn Meal, bag (•)	1
Detergent Powder (with phosphates) (•)	1
Detergent Powder (without phosphates) (•)	1
Flour, bag (•)	1
Fur (from a pet animal) (•)	15
Hair (from a student) (•)	15
Medicine Droppers, pkg. of 12 (17 M 0230)	2

Red Food Color (or ketchup, etc.) (•)	1
Salt (sodium chloride), 500 g (37 M 5480)	1
Sand, 500 g (20 M 7423)	1

Investigation 2–1 (pp. 81–86)

Biological Supplies

Seeds, Bean (untreated) (86 M 8016)	1

Lab Equipment

Petri Dishes, disposable, pkg. of 20 (18 M 7101)	3

Miscellaneous

Paper Towels, roll (15 M 9844)	1

Investigation 2–2 (pp. 87–90)

Biological Supplies

Chiton (preserved), pkg. of 10 (68 M 7002)	2
Daphnia magna (live) (87 M 5210)	2
Euglena sp. (live) (87 M 0100)	2
Nostoc (live), ball colonies (86 M 2155)	2
Saccharomyces cerevisiae, culture (live) (85 M 5000)	15
Tapeworm, adult (preserved) (68 M 0551)	15

Lab Equipment

Container (15 M 0539)	30
Coverslips, pkg. of 100 (14 M 3555)	1
Microscope, Compound Light (24 M 2310)	15
Microscope Slides, pkg. of 72 (14 M 3500)	1

Miscellaneous

Medicine Droppers, pkg. of 12 (17 M 0230)	2
Paper Towels, roll (15 M 9844)	1
Ruler, metric (14 M 0810)	30

Investigation 3–1 (pp. 91–94)

Biological Supplies

Amoeba culture (live), vital stained (87 M 0380)	1
Elodea (live), pkg. of 10 (86 M 7500)	2

Lab Equipment

Coverslips, pkg. of 100 (14 M 3555)	1
Microscope, Compound Light (24 M 2310)	15
Microscope Slides, pkg. of 72 (14 M 3500)	1

Miscellaneous

Medicine Droppers, pkg. of 12 (17 M 0230)	2

Investigation 3–2 (pp. 95–98)

Biological Supplies

Frog Blood, slide (92 M 3640)	15
Frog Skin, slide (92 M 3643)	15
Frog Sperm, slide (92 M 8803)	15

Lab Equipment

Microscope, Compound Light (24 M 2310)	15

Miscellaneous

Lens Paper, box of 50 (15 M 8250)	15

Enrichment

Cell Structure and Function (color slides) (170 M 9109)	1
The Cell (transparency set) (75 M 0110)	1

Investigation 4–1 (pp. 99–102)

Lab Equipment

Balance, Triple Beam (15 M 6057)	15
Beakers, 250 mL (17 M 4040)	24
Beakers, 600 mL (17 M 4060)	30

Miscellaneous

Corn Syrup, 500 mL (39 M 1461)	15
Eggs, fresh, 12	3
Labels, pkg. of 240 (15 M 1832)	1
Vinegar (acetic acid solution), 1 pt. (39 M 0138)	2

Investigation 5–1 (pp. 103–106)

Biological Supplies

Elodea (live), pkg. of 10 (86 M 7500)	2
Snails, Pond (live), pkg. of 12 (87 M 4141)	5

Chemicals/Media

Bromothymol Blue, 0.1%, 500 mL (38 M 9100)	1

Lab Equipment

Culture Tube and Cap, 16 × 150 mm (17 M 1340)	120
Dechlorinator (21 M 2292)	1
Gloves, disposable, box of 100 (15 M 1071)	1
Light Bulb (150 W, for light source) (36 M 4173)	15
Light Source (36 M 4168)	15

Personal Safety Set
    antifog goggles
    cotton-lined latex gloves
    vinyl apron (15 M 3041)
	30

Miscellaneous

Paper, white (•)	15
Wax Pencil (15 M 1155)	15

Investigation 5–2 (pp. 107–112)

Chemicals/Media

Limewater (calcium hydroxide solution), 4 L (37 M 2631)	1

Lab Equipment

Beakers, 250 mL (17 M 4040)	36
Bottles, Vacuum (insulated) (•)	30
Erlenmeyer Flask, 250 mL (17 M 2982)	7
Glass Tubing (17 M 0931)	1
Gloves, disposable, box of 100 (15 M 1071)	1

Personal Safety Set
    antifog goggles
    cotton-lined latex gloves
    vinyl apron (15 M 3041)
	30

Rubber Tubing (50') (15 M 1138)	1
Stoppers, 2-hole, size 9, pkg. of 10 (15 M 8519)	3
Thermometer (15 M 1462)	15

Miscellaneous

Molasses, 350 mL (39 M 2298)	3
Paper, Graph (15 M 3835)	2
Sucrose, 500 g (39 M 3182)	3
Yeast, Baker's (dry), 100 g (38 M 5826)	1

Investigation 6–1 (pp. 113–116)

Miscellaneous

Paper, white (•)	30
Pencils, No. 2, pkg. of 12 (15 M 9816)	3

Investigation 6–2 (pp. 117–122)

Miscellaneous

Paper, white (•)	30
Pencils, No. 2, pkg. of 12 (15 M 9816)	3

Investigation 7–1 (pp. 123–128)

Lab Equipment

Scissors (14 M 0988)	15

Miscellaneous

Index Cards, pkg. of 1,000 (15 M 9807)	1
Marker Set, 4 colors (15 M 4635)	15
Paper Clips, pkg. of 1,000 (15 M 9815)	1
Paper Grocery Bags (•)	30
Tape, cellophane, roll (15 M 1957)	1

Investigation 8–1 (pp. 128–132)

Lab Equipment

Scissors (14 M 0988)	15

Miscellaneous

Paper, white (•)	15
Tape, cellophane, roll (15 M 1957)	15

Enrichment

Chromosome Spread: Down Syndrome Male (33 M 1048)	1
Chromosome Spread: Down Syndrome Female (33 M 1049)	1
Chromosome Spread: Klinefelter's Syndrome (33 M 1050)	1
Chromosome Spread: Turner's Syndrome (33 M 1051)	1
Human Karyotype Form (33 M 1045)	1
Landmarks of the Human Genome, Macintosh® (74 M 3673)	1
Landmarks of the Human Genome, IBM® (74 M 3674)	1

Investigation 8–2 (pp. 133–136)

Biological Supplies

Euglena sp. (live) (87 M 0100)	15

Chemicals/Media

Detain™ (37 M 7950)	15
Euglena Medium, 1 L (88 M 5200)	4

Lab Equipment

Beaker, 100 mL (17 M 4020)	45
Coverslips, pkg. of 100 (14 M 3555)	10
Graduated Cylinder, 10 mL (17 M 0170)	15
Incubator (15 M 0060)	1
Microscope Slides, pkg. of 72 (14 M 3500)	1
Microscope, Compound Light (24 M 2310)	15

Miscellaneous

Medicine Droppers, pkg. of 12 (17 M 0230)	4
Toothpicks, pkg. of 750 (15 M 9864)	1
Wax Pencil (15 M 1155)	30

Investigation 9-1 (pp. 137-142)

Miscellaneous

Paper, Graph (15 M 3835) 1
Pencils, colored, pkg. of 12 (15 M 2576) 15

Investigation 10-1 (pp. 143-150)

No additional materials besides the textbook and lab manual are needed for this investigation.

Investigation 11-1 (pp. 151-156)

Miscellaneous

Calculator (27 M 3055) 15
Protractor (15 M 4067) 30
Ruler, metric (14 M 0810) 30

Investigation 12-1 (pp. 157-162)

Lab Equipment

Meter Stick (15 M 4065) 15

Miscellaneous

Glue, white, 4 oz. (15 M 9806) 15
Marker Set, 4 colors (15 M 4635) 15
Poster Board (18M 2105) 15
Stakes, pkg. of 12 (21 M 0364) 5
Sticker Tag Set, 8 colors (15 M 9805) 15
String (15 M 9863) 15

Investigation 13-1 (pp. 163-168)

Biological Supplies

Spirogyra culture (live—teacher must subculture) (86 M 0658) 1

Chemicals/Media

Algae Culture Solution (makes 1 L) (88 M 3250) 1

Lab Equipment

Aquarium/Terrarium, 1 gal. (21 M 2101) 45
Balance, Triple Beam (15 M 6057) 15
Dechlorinator (21 M 2292) 1
Fluorescent Light (bulbs included) (20 M 5100) 1
Gloves, disposable, box of 100 (15 M 1071) 1
Personal Safety Set
  antifog goggles
  cotton-lined latex gloves
  vinyl apron (15 M 3041) 30

Miscellaneous

Detergent Powder (with phosphates) (15 M 1287) 1
Detergent Powder (without phosphates) (15 M 1288) 1
Paper Towels, roll (15 M 9844) 1
Plastic Wrap, roll (15 M 9858) 1
Wax Pencil (15 M 1155) 30

Investigation 14-1 (pp. 169-172)

Biological Supplies

Seeds, Radish (86 M 8280) 8

Chemicals/Media

Hydrochloric Acid, 0.1 M, 1 L (37 M 9561) 1
pH Indicator Strips, pkg. of 200 (15 M 2505) 2

Lab Equipment

Beakers, 50 mL (17 M 4010) 60
Filter Papers, 9 cm, pkg. of 100 (15 M 2835) 1
Graduated Cylinder, 10 mL (17 M 0170) 15
Petri Dishes, disposable, pkg. of 20 (18 M 7101) 2

Miscellaneous

Apple, fresh red (•) 15
Cotton, 1 lb. (15 M 3830) 3
Steel Wool, 16 pads (15 M 8798) 1
Wax Pencil (15 M 1155) 30

Investigation 14-2 (pp. 173-178)

Biological Supplies

Oil-Degrading Microbes Set (live) (85 M 7000) 1

Chemicals/Media

Microbial Growth Measurement Strips, pkg. of 25 (15 M 1996) 2
Refined Oil, 4 oz. (750 M 3502) 1
Water, distilled (88 M 7005) 1

Lab Equipment

Culture Tube and Cap, 16×150 mm (17 M 1340) 30
Test Tube Rack (18 M 0010) 15

Investigation 15-1 (pp. 179-182)

For this lab, gather 15 sets of 20 common items, such as leaves, acorns, twigs, rocks, pine needle bundles, pine cones, paper clips, nails, screws, soil, chalk, coins, toothpicks, thumbtacks, etc.

Investigation 16-1 (pp. 183-186)

Biological Supplies

Bacillus subtilis (live), with growth media (85 M 1840) 1
E. coli (live), with growth media (85 M 1860) 1

Chemicals/Media

Antibiotic Discs, control, pkg. of 100 (38 M 1600) 1
Antibiotic Discs, Erythromycin, pkg. of 50 (38 M 1605) 1
Antibiotic Discs, Kanamycin, pkg. of 50 (38 M 1606) 1
Antibiotic Discs, Tetracycline, pkg. of 50 (38 M 1611) 1
Plates, Nutrient Agar, pkg. of 6 (88 M 0905) 6

Lab Equipment

Forceps (14 M 0999) 15
Gloves, disposable, box of 100 (15 M 1071) 1
Incubator (15 M 0060) 1
Personal Safety Set
  antifog goggles
  cotton-lined latex gloves
  vinyl apron (15 M 3041) 30
Swabs, sterile, pkg. of 100 (14 M 5502) 3

Miscellaneous

Ruler, metric (14 M 0810) 30
Wax Pencil (15 M 1155) 30

Investigation 16-2 (pp. 187-190)

Chemicals/Media

Plates, Nutrient Agar, pkg. of 6 (88 M 0905) 13
Soap, Antibacterial with Iodine, 16 oz. (15 M 9829) 3

Lab Equipment

Incubator (15 M 0060) 1

Miscellaneous

Scrub Brush (for hands) (15 M 9106) 15
Soap, bar (•) 15
Tape, cellophane, roll (15 M 1957) 1
Wax Pencil (15 M 1155) 30

Investigation 17-1 (pp. 191-196)

Biological Supplies

Pond Water, Dry Mix (87 M 9055) 1

Lab Equipment

Coverslips, pkg. of 100 (14 M 3555) 1
Culture Dish (17 M 0560) 15
Microscope, Compound Light (24 M 2310) 15
Microscope Slides, pkg. of 72 (14 M 3500) 1

Miscellaneous

Medicine Droppers, pkg. of 12 (17 M 0230) 2
Wax Pencil (15 M 1155) 30

Enrichment

Dichotomous Key to Pond Microlife, Apple II (74 M 1858) 1
Dichotomous Key to Pond Microlife, Mac (74 M 0041) 1

Investigation 17–2 (pp. 197–200)

Biological Supplies

Anabaena culture (live) (86 M 1800) 1
Chlorella culture (live) (86 M 0126) 1
Nostoc culture, ball colonies (live) (86 M 2155) 1

Chemicals/Media

Lugol's Iodine Solution (39 M 1685) 1
Starch, soluble, 100 g (39 M 3275) 1

Lab Equipment

Coverslips, pkg. of 100 (14 M 3555) 1
Gloves, disposable, box of 100 (15 M 1071) 1
Microscope, Compound Light (24 M 2310) 15
Microscope Slides, pkg. of 72 (14 M 3500) 1
Personal Safety Set
   antifog goggles
   cotton-lined latex gloves
   vinyl apron (15 M 3041) 30
Spot Plate (15 M 3980) 15
Watch Glass, Syracuse, pkg. of 12 (17 M 0530) 4

Miscellaneous

Lens Paper, box of 50 (15 M 8250) 15
Medicine Droppers, pkg. of 12 (17 M 0230) 2
Wax Pencil (15 M 1155) 30

Investigation 18-1 (pp. 201–206)

Biological Supplies

Seeds, Sunflower (86 M 8320) 15
Seeds, Tomato (86 M 8340) 15

Chemicals/Media

Mineral Solutions (266 M 8400) 1
Water, distilled (88 M 7005) 1

Lab Equipment

Germination Tray (20 M 3230) 15
Gloves, disposable, box of 100 (15 M 1071) 1
Personal Safety Set
   antifog goggles
   cotton-lined latex gloves
   vinyl apron (15 M 3041) 30

Miscellaneous

Containers with Lids, 1 qt. (18 M 9919) 10
Corks, pkg. of 100 (15 M 8364) 2
Cotton, 1 lb. (15 M 3830) 3
Ruler, metric (14 M 0810) 15
Vermiculite, 14 lb. bag (20 M 8630) 1

Enrichment

Plant Mineral Deficiency Activity (20 M 8400) 15

Investigation 19–1 (pp. 207–212)

Biological Supplies

Aquatic Plants, floating (live), jar (86 M 7650) 1

Chemicals/Media

Cobalt Chloride Test Paper, 12 sheets (37 M 1340) 1

Lab Equipment

Coverslips, pkg. of 100 (14 M 3555) 1
Microscope, Compound Light (24 M 2310) 15
Microscope Slides, pkg. of 72 (14 M 3500) 1
Scalpel (14 M 0705) 15

Miscellaneous

Medicine Droppers, pkg. of 12 (17 M 0230) 2
Nail Polish, clear (37 M 1862) 15
Paper Clips, pkg. of 1,000 (15 M 9815) 1
Plastic Bags, zipper type, pkg. of 10 (18 M 6924) 27
Tape, cellophane, roll (15 M 1957) 15
Terrestrial Plant (Geranium) (86 M 6900) 15
Twist Ties, 250' roll (14 M 0947) 1
Watch with second hand (15 M 0512) 15

Investigation 20-1 (pp. 213–216)

Chemicals/Media

Ascorbic Acid, 100 g (39 M 2006) 1
Indophenol Solution, 0.1%, 500 mL (37 M 9542) 1

Lab Equipment

Gloves, disposable, box of 100 (15 M 1071) 1
Hot Plate (15 M 8055) 15
Personal Safety Set
   antifog goggles
   cotton-lined latex gloves
   vinyl apron (15 M 3041) 30
Test Tube, 13 × 100 mm (17 M 0620) 165
Test Tube Rack (18 M 0010) 15

Miscellaneous

Juice, assorted varieties (•)   enough for 15 lab groups

Investigation 20-2 (pp. 217–220)

Biological Supplies

Seeds, Lima Bean, pkg. of 50 (86 M 8008) 1

Chemicals/Media

Gibberellic Acid, 5 g (39 M 1446) 1
Indoleacetic Acid, 5 g (39 M 1661) 1

Lab Equipment

Erlenmeyer Flask, 250 mL (17 M 2982) 75
Germination Tray (20 M 3230) 15
Gloves, disposable, box of 100 (15 M 1071) 1
Personal Safety Set
   antifog goggles
   cotton-lined latex gloves
   vinyl apron (15 M 3041) 30
Scalpel (14 M 0705) 15

Miscellaneous

Cotton, 1 lb. (15 M 3830) 1
Labels, pkg. of 240 (15 M 1832) 1
Ruler, metric (14 M 0810) 30
Trowel (20 M 7015) 15
Vermiculite, 14 lb. bag (20 M 8630) 1
Wax Pencil (15 M 1155) 30

Investigation 21-1 (pp. 221–228)

Biological Supplies

Earthworms (live), pkg. of 10 (87 M 4660) 2
Eel, Vinegar, culture (live) (87 M 2900) 1

Lab Equipment

Beakers, 600 mL (17 M 4060) 15
Coverslips, 100 (14 M 3555) 1
Culture Dish (17 M 0560) 15
Dissection Pan Set (14 M 7011) 15
Hand Lens (25 M 1350) 15
Light Bulb (150 W, for light source) (36 M 4173) 15
Light Source (36 M 4168) 15
Microscope, Compound Light (24 M 2310) 15
Microscope Slides, concave, pkg. of 12 (14 M 3510) 2
Stereomicroscope (24 M 4602) 15

Miscellaneous

Construction Paper, 50 sheets (15 M 9841) 1
Ice (crushed) (•)
Medicine Droppers, pkg. of 12 (17 M 0230) 2
Paper Towels, roll (15 M 9844) 1
Tape, cellophane, roll (15 M 1957) 30
Vinegar, Cider, 1 pt. (39 M 0139) 1

Investigation 22-1 (pp. 229–236)

Chemicals/Media

ATP Powder, 1 g (39 M 0174) 1
Firefly Lantern Extract, 50 mg (37 M 1580) 1
Water, distilled (88 M 7005) 1

Miscellaneous

Cloth (•) 1
Dark Fabric (15 M 2539) 1
Insects (•) 15
Jars, clear plastic, with lid (18 M 7192) 30
Needle, Sewing (•) 15
Pot, 3 in., pkg. of 10 (20 M 2130) 3
Potting Soil, 8 lb. bag (20 M 8306) 2
Screen, metal window type, 1 yd. (15 M 0002) 15
Thread, spool, pkg. of 3 (15 M 9837) 5

Investigation 23–1 (pp. 237–242)

Biological Supplies
Snails, Land (live), pkg. of 6 (87 M 4306)   3
Snails, Pond (live), pkg. of 12 (87 M 4141)   2

Lab Equipment
Aquarium/Terrarium, 1 gal (21 M 2101)   30
Hand Lens (25 M 1350)   15
Light Bulb (150 W, for light source) (36 M 4173)   15
Light Source (36 M 4168)   15
Petri Dishes, disposable, pkg. of 20 (18 M 7101)   1
Stereomicroscope (24 M 4602)   15

Miscellaneous
Box (shoe box, etc.) (•)   15
Pencils, No. 2, pkg. of 12 (15 M 9816)   2

Investigation 24–1 (pp. 243–250)

Biological Supplies
Crickets (live), pkg. of 50 (87 M 6102)   1

Lab Equipment
Beakers, 600 mL (17 M 4060)   45
Ice Bucket (18 M 1625)   15
Light Bulb (150 W, for light source) (36 M 1473)   15
Light Source (36 M 4168)   15
Stereomicroscope (24 M 4602)   15

Miscellaneous
Apple, fresh red (•)   15
Box (shoe box, etc.) (•)   15
Cloth Square, 35 × 35 cm (15 M 2538)   15
Ice (crushed) (•)
Plastic Bags, zipper type (18 M 6924)   3
Screening (15 M 0002)   15
Tape, masking, roll (15 M 9828)   5

Investigation 24–2 (pp. 251–258)

Biological Supplies
Centipede (preserved), pkg. of 10 (68 M 3112)   2
Crayfish (live), pkg. of 12 (87 M 6031)   2
Crickets (live), pkg. of 50 (87 M 6102)   1
Millipedes (live), pkg. of 12 (87 M 6971)   2
Pillbugs (live), pkg. of 45 (87 M 5520)   1
Tarantula (live) (87 M 6965)   1
Tarantula (preserved) (68 M 3071)   15

Lab Equipment
Aquarium/Terrarium, 1 gal (21 M 2101)   30
Beakers, 600 mL, pkg. of 6 (17 M 4060)   3
Dechlorinator (21 M 2292)   2
Probe (14 M 0958)   15
Watch Glass, Syracuse, pkg. of 12 (17 M 0530)   2

Miscellaneous
Paper Towels, roll (15 M 9844)   1
Screening (15 M 0002)   15

Investigation 25–1 (pp. 259–268)

Biological Supplies
Carp (live) (87 M 8103)   5
Fish Scale Slide Set (95 M 0825)   15
Goldfish (live), pkg. of 12 (87 M 8100)   2

Lab Equipment
Beakers, 600 mL (17 M 4060)   15
Dechlorinator (21 M 2292)   2
Fish Net (21 M 3702)   15
Flashlight (pen-size) (15 M 4000)   15
Forceps (14 M 0999)   15
Stereomicroscope (24 M 4602)   15

Miscellaneous
Box (shoe box, etc.) (•)   15
Medicine Droppers, pkg. of 12 (17 M 0230)   2
Microscope Slides, pkg. of 72 (14 M 3500)   1

Investigation 26–1 (pp. 269–274)

Miscellaneous
Pencils, No. 2, pkg. of 12 (15 M 9816)   2

Enrichment
Field Guide to Birds, Eastern Region (32 M 2103)   15
Field Guide to Birds, Western Region (32 M 2104)   15

Investigation 27–1 (pp. 275–278)

Chemicals/Media
Ethyl Alcohol, 95% (39 M 0283)   2
Soap, Antibacterial with Iodine, 16 oz. (15 M 9829)   2

Lab Equipment
Gloves, disposable, box of 100 (15 M 1071)   1
Light Bulb (150 W, for light source) (36 M 4173)   15
Light Source (36 M 4168)   15
Personal Safety Set
    antifog goggles
    cotton-lined latex gloves
    vinyl apron (15 M 3041)   30
Swabs, sterile, pkg. of 100 (14 M 5502)   1

Miscellaneous
Paper Towels, roll (15 M 9844)   1
Paper, erasable bond (•)   5
Ruler, metric (14 M 0810)   30

Investigation 28–1 (pp. 279–284)

Chemicals/Media
Quinine Sulfate Solution, 0.1%, 500 mL (37 M 9545)   1

Lab Equipment
Glass Rods, pkg. of 10 (17 M 6010)   2
Graduated Cylinder, 25 mL (17 M 0171)   15
Swabs, sterile (14 M 5502)   3

Miscellaneous
Apple, fresh red (•)   15
Marker Set, 4 colors (15 M 4635)   15
Onion, fresh (•)   15
Plastic Cups, pkg. of 25 (18 M 3675)   5
Potato, fresh (•)   15
Salt (Sodium Chloride), 500 g (37 M 5480)   1
Spoon, plastic (•)   75
Sucrose, 500 g (39 M 3182)   1
Vinegar (acetic acid solution), 1 pt. (39 M 0138)   1

Investigation 29–1 (pp. 285–288)

Biological Supplies
Daphnia magna (live) (87 M 5210)   1
Chemicals/Media
Adrenalin Solution, 100 mL (38 M 2053)   1

Lab Equipment
Gloves, disposable, box of 100 (15 M 1071)   1
Microscope, Compound Light (24 M 2310)   15
Microscope Slides, concave, pkg. of 12 (14 M 3510)   2
Personal Safety Set
    antifog goggles
    cotton-lined latex gloves
    vinyl apron (15 M 3041)   30
Petri Dishes, disposable, pkg. of 20 (18 M 7101)   1
Pipets, transfer, pkg. of 500 (18 M 2971)   1

Miscellaneous
Paper Towels, roll (15 M 9844)   1
Spoon, plastic (•)   15
Watch with second hand (15 M 0512)   15

Investigation 30–1   (pp. 289–292)

Biological Supplies
Daphnia magna (live) (87 M 5210) — 1

Chemicals/Media
Acetylcholine Solution, 100 mL (38 M 2021) — 1
Ethyl Alcohol, 95% (39 M 0283) — 1
Nicotine Solution, 100 mL (39 M 2457) — 1

Lab Equipment
Beaker, 100 mL (17 M 4020) — 15
Coverslips, pkg. of 100 (14 M 3555) — 1
Gloves, disposable, box of 100 (15 M 1071) — 1
Microscope, Compound Light (24 M 2310) — 15
Microscope Slides, Concave, pkg. of 12 (14 M 3510) — 2
Personal Safety Set
    antifog goggles
    cotton-lined latex gloves
    vinyl apron (15 M 3041) — 30

Miscellaneous
Antacid, bottle of 100 (•) — 1
Aspirin, bottle of 100 (•) — 1
Coffee (1 cup) (•) — 1
Cough Drops, package (•) — 1
Medicine Droppers, pkg. of 12 (17 M 0230) — 13
Nasal Spray (•) — 1
Paper Towels, roll (15 M 9844) — 1
Tea, brewed (1 cup) (•) — 1
Watch with second hand (15 M 0512) — 15

Investigation 31–1   (pp. 293–296)

Biological Supplies
Blood Cells (normal), slide (93 M 6540) — 15
Blood Cells (sickle-cell), slide (93 M 8120) — 15

Lab Equipment
Microscope, Compound Light (24 M 2310) — 15

Miscellaneous
Pencils, colored, pkg. of 12 (15 M 2576) — 15
Ruler, metric (14 M 0810) — 30

Investigation 32–1   (pp. 297–300)

Chemicals/Media
Hydrochloric Acid, 0.1 M, 1 L (37 M 9561) — 1
Phenol Red, 5 g (38 M 9431) — 1

Lab Equipment
Dropper Bottle, 1 oz. (17 M 1451) — 30
Test Tube, 15 × 125 mm (17 M 0620) — 30

Investigation 33–1   (pp. 301–306)

Chemicals/Media
Amylase, 20 g (38 M 0058) — 1
Hydrochloric Acid, 0.1 M, 1 L (37 M 9561) — 2
Litmus Paper, blue, pkg. of 1,200 (15 M 3105) — 1
Litmus Paper, red, pkg. of 1,200 (15 M 3107) — 1
Lugol's Iodine Solution (39 M 1685) — 1
Pancreatin, 100 g (30 M 2728) — 1
Pepsin, 100 g (39 M 2866) — 1
Soap, Antibacterial with Iodine, 16 oz. (15 M 9829) — 3
Sodium Hydroxide, dilute, 1 L (37 M 9546) — 1
Starch, soluble, 100 g (39 M 3275) — 1

Lab Equipment
Beakers, 600 mL (17 M 4060) — 30
Gloves, disposable, box of 100 (15 M 1071) — 1
Graduated Cylinder, 10 mL (17 M 0170) — 15
Hot Plate (15 M 8055) — 15
Personal Safety Set
    antifog goggles
    cotton-lined latex gloves
    vinyl apron (15 M 3041) — 30
Scalpel (14 M 0705) — 15
Stoppers, pkg. of 110 (15 M 8459) — 2
Test Tube, 15 × 125 mm (17 M 0620) — 150
Test Tube Rack (18 M 0010) — 15
Thermometer (15 M 1462) — 15

Miscellaneous
Eggs, fresh, 12 (•) — 1
Ice (•)
Vegetable Oil, 500 mL (37 M 9539) — 1
Wax Pencil (15 M 1155) — 15

Investigation 33–2   (pp. 307–310)

Chemicals/Media
Benedict's Solution (37 M 0698) — 1
Glucose, 500 g (39 M 1357) — 1
Glucose Test Paper, pkg. of 50 strips (14 M 4107) — 1
Lugol's Iodine Solution (39 M 1685) — 1

Lab Equipment
Beaker, 250 mL (17 M 4040) — 30
Hot Plate (15 M 8055) — 15
Microscope Slides, concave, pkg. of 12 (14 M 3510) — 2
Test Tube, 15 × 125 mm (17 M 0620) — 30

Miscellaneous
Lactose-Intolerance Product (liquid) (39 M 0140) — 2
Medicine Droppers, pkg. of 12 (17 M 0230) — 4
Milk, 1 pt. (•) — 15
Toothpicks, pkg. of 750 (15 M 9864) — 1

Investigation 34–1   (pp. 311–316)

Biological Supplies
Chick Embryo Slide, 28-Hour (92 M 9035) — 15
Chick Embryo Slide, 33-Hour (92 M 9040) — 15
Chick Embryo Slide, 43-Hour (92 M 9058) — 15
Chick Embryo Slide, 56-Hour (92 M 9070) — 15
Sea Urchin Slide (all stages) (92 M 8330) — 15

Lab Equipment
Microscope, Compound Light (24 M 2310) — 15
Stereomicroscope (24 M 4602) — 15

Name _____ Date _____ Class _____

Imagining Solutions: Problem Solving

This activity gives students an opportunity to use their imagination while comparing materials used to store solar energy. The activity can be extended, making it as open-ended as you wish. Allow the students to select the substances to be tested, but safety, cost, and availability should be considered.

Background

Imagination can be as important to the biologist as it is to a writer or story-teller. For this activity, you will use your imagination to suggest solutions to a problem involving heat storage. Problems such as the one you are about to encounter must be solved to make the use of solar energy practical.

Objectives

In this activity you will:
• **imagine** solutions to a problem

Materials

• heat-absorbing materials (3, your choice)
• beakers (3), 400-mL
• plastic wrap
• wax pencil
• thermometer °C
• colored pencils (3)

Preparation

Water, air, cooking oil, salt, sugar, starch, and gravel are materials to consider. Avoid materials that are toxic or inflammable.

If possible, have the students measure the temperatures of the beakers' contents first thing in the morning, during class, and at the end of the day. This work could be divided among members of the group.

1. Form cooperative groups of three students and consider the problem of heat storage. Propose three materials that could be used to collect and store heat. Take into consideration safety, availability, cost, and ease of use.
2. Have your teacher approve your proposed materials.
3. Make a table similar to the one shown below for recording your data.

Day	After 24 Hours in the:	Temperatures of Materials Being Tested		
1	Dark			
2	Sun			
3	Dark			

Procedure

Introduce the activity with a discussion as to why students think it is important to use the imagination. Guide students to develop a relationship between using the imagination to solve problems and technological advances.

4. Fill each beaker with one of the materials to be tested. Cover each beaker with plastic wrap and label accordingly with a wax pencil.
5. Set the beakers in a dark place for 24 hours.
6. Remove the plastic wrap to measure the temperature of the material in each beaker, and record your results.
7. Re-cover the beakers with plastic wrap and transfer to a sunny location, such as the window sill, for 24 hours.
8. Remove the plastic wrap to measure the temperature of the material in each beaker, and record your results.

9. Re-cover the beakers and return to the dark location for 24 hours.
10. Measure the temperature of the material in each beaker and record your results.
11. Use your colored pencils to make a line graph that compares the temperatures of the three beakers over time. Use a different color line to represent each of the beakers.
12. Clean up your materials and wash your hands before leaving the lab.

Analysis

1. **Summarizing Data** Summarize your findings.

   Answers will vary depending on the materials being tested. The greatest fluctuation in temperature will probably occur in air.

2. **Analyzing Methods** At what point does imagination enter into the problem-solving process?

   Imagination entered into the problem-solving process at the very beginning, when hypotheses are formulated.

3. **Evaluating Methods** Why might imagination be one of the most important traits of the biologist?

   Imagination can lead to a hypothesis that can be tested. Without a hypothesis, the scientific process cannot continue.

Comparing Living and Nonliving Things

For many students, this may be their first exposure to "life" in a biological sense. They may associate life only with movement in animals and green color in plants. This activity should give the students an opportunity to explore a variety of life forms. The correct answer is not as important as showing evidence of observation and analysis. This activity is meant to initiate the learning cycle.

Background

Do all living things share certain traits? If so, what might these traits be? In this activity, you will try to answer these questions.

Objectives

In this activity you will:
• *observe* and *compare* living and nonliving things

Materials

• unlabeled specimens
• stereomicroscopes
• compound microscopes

Preparation

1. Form cooperative teams of two students.
2. Make a table similar to the one shown below.

Trait	Specimen									
	1	2	3	4	5	6	7	8	9	10
1.										
2.										
3.										
4.										
5.										
6.										
7.										
8.										
9.										
10.										

3. In the first column of the table, list 10 traits or processes that you associate with living organisms.

Procedure

Let the students know that the emphasis of the activity is on observation and making comparisons. The traits identified at the beginning of the activity are a basis for future work.

4. Closely observe a specimen. Follow any special directions for viewing the specimen that your teacher may supply.
5. Place an "X" in your data table for each trait you observe in the specimen.
6. Add to your table any new traits that come to mind during your observations.
7. Repeat steps 4–6 for each of the specimens.

Analysis

1. **Summarizing Data** Which specimens are alive? dead? nonliving?

   Answers will vary according to the specimens observed.

2. **Comparing Groups** What traits do living things have in common?

   Answers will vary. Some examples might include movement, eating, response to light or other stimuli, vocalization, excretion or elimination of waste, reproduction, growth, repair, or energy consumption.

3. **Comparing Groups** What traits do nonliving things have in common?

   Answers will vary but generally are the converse of traits of living organisms.

4. **Evaluating Methods** Why can it be difficult to observe all the traits common to living things?

   Answers will vary but should refer to temporal and perceptual limitations.

Place a variety of numbered specimens at different stations around the lab. Provide supplemental instructions whenever necessary. For example, instructions for using the microscope, moving the specimen, or removing a cover might be necessary. Select a wide variety of living, nonliving, and dead specimens. Specimens might include:

- unprocessed yogurt, include a slide for use under high power
- blue cheese or bread mold, include wet-mount slides
- dry mung beans or other seeds
- dried or activated yeast, show under high power
- pumice, clean sand, or powdered clay
- powdered charcoal, show under high power
- coral, sea shells, sponge, or snail shell
- dead insect
- living insect, crayfish, or earthworm
- goldfish or minnows

- cork, whole or thin-section under high power
- lichen on a rock, including a prepared slide under low power
- fossil
- loofa (Luffa) "sponge"—actually a dried plant stem available in health food stores
- synthetic sponge
- potted plant
- protozoans or algae collected from a stagnant pond or slow-moving stream
- slime mold growing on a moist paper towel in a petri dish or on plain agar in a petri dish

Modeling Cells: Surface Area to Volume

Surface area-to-volume relationships play a role in many aspects of biology. In this activity, the students build simple models to investigate why and how this ratio might limit the size of organisms.

Background

Are there limits as to how large organisms can grow? Some humans are very tall but do not grow as large as trees. Some large insects do exist but never grow to reach the sizes you might see in science fiction films. Why? One reason is that with increasing height there is a disproportionate increase in volume (or weight). If the height of an elephant were doubled, its weight would increase by eight times its original weight. An elephant cannot grow larger, because its legs could not support the increase in weight.

In this activity you will examine surface area-to-volume ratios on a small scale, using some model cells. You will use the collected data to reach some conclusions as to why this ratio might limit the size of a cell.

Objectives

In this activity you will:
- **construct** and **analyze** various cell models
- **measure** volume and calculate surface area
- **calculate** surface area-to-volume ratios
- **form conclusions** about size limitations, using your data

Materials

- scissors
- cell model cutouts (3)
- poster board
- tape
- metric ruler
- sand
- funnel
- large graduated cylinder
- calculator (optional)

Preparation

1. Form cooperative groups of two students. Cut out three cell models and fold each to form a three-dimensional shape. Cell dimensions should be recorded in the table in step 2 below. Use tape where directed so that the models hold their shape.

Procedure

2. Using the metric ruler, measure the length, width, and height dimensions of each model. Record the dimensions in a table like the one shown below.

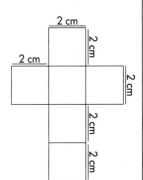

Cell	Dimensions (cm)	Surface Area (cm^2)	Volume (cm^3)	Ratio Surface Area: Volume
A	2 x 2	24	8	3
B	4 x 4	96	64	1.5
C	8 x 8	384	512	0.75

Graph paper can be used to construct the cell models. This facilitates drawing straight lines and right angles. If necessary, each team can be given pre-drawn patterns for each of the three cell models. Also, you might wish to select a range of measurements.

Have the class speculate on the factors that limit cells to microscopic sizes. Look for reasonable and logical responses rather than the correct answer. Discuss why models can be used to investigate some phenomena.

3. Fill each model with sand. Level off the sand at the top of the model, using the ruler.
4. Find the volume of sand in each model. You can do this two ways.
   a. Measure the amount of sand in each model, using a graduated cylinder. Pour the sand through the funnel into the graduated cylinder or a measuring cup.
      One millimeter = one cubic centimeter (1 cm^3).
   b. Calculate the volume using the formula in step 2 of Analysis (below).

Analysis

1. **Analyzing Models** Complete the Data Table by calculating the area and volume of each model. To calculate total surface area for each model, find the area of each side (length x width) then multiply that number by 6. Enter the data in your table. Why do you need to multiply by 6?

   To calculate the total surface area, measures of the four sides, bottom, and top must be
   considered, making a total of six.

2. **Analyzing Models** To calculate the volume of sand, use the following formula:

$$volume = length \times width \times height$$

   Record the volume of each model in your Table.

3. **Analyzing Models** Calculate the surface area-to-volume ratio for each model. Use the following formula:

$$\frac{surface\ area}{volume} = ratio$$

   Record the value in your table.

4. **Comparing and Contrasting Models** Which model has the largest surface area?

   Model C

   Which model has the largest volume?

   Model C

   Which model has the largest ratio?

   Model A

5. **Making Predictions** To maintain life, materials must be able to move into and out of a cell. What might be the advantage of having a large surface area?

   The cell could take in more materials.

   What might be the disadvantage of having a large volume?

   If the cell was larger, fewer of its interior structures would be near the cell membrane,
   making the cell less efficient.

Demonstrating Diffusion

While the demonstration of diffusion is the primary objective of this activity, the relationship between temperature and diffusion is used to rouse the student's curiosity. Although the diffusion of food coloring through a beaker of water is used as an example in many textbooks, many students have never actually seen it firsthand.

Background

Molecules must move about in a cell in order for a cell to survive. In this activity, you will see the movement of molecules and study how temperature affects this movement.

Objectives

In this activity you will:
- *observe* diffusion
- *recognize* the relationship between temperature and the rate of diffusion

Materials

- beakers (3), 400-mL
- wax pencil
- water
- ice cubes
- hot plate
- insulated glove
- hot pad (optional)
- thermometer
- food coloring
- timer

Preparation

1. Form cooperative groups of two students. Complete steps 2–10.
2. Label three clean 400-mL beakers,*Cold*, *Medium*, and *Hot*, respectively.

Procedure

3. Add cool or lukewarm tap water to each beaker until it is approximately half full.
4. Place two ice cubes in the Cold beaker. Set both the Cold beaker and the Medium beaker aside.

5. **CAUTION: A hot plate's high temperature can cause injury.** Place the Hot beaker on a hot plate and heat the water to boiling.
6. **CAUTION: Boiling water can cause injury.** Use an insulated glove to carefully remove the beaker from the hot plate. Use the hot pad when placing hot objects on surfaces that are not suited for high temperatures.
7. Use a thermometer to find the initial temperature of the water in each beaker. Record your results in a table similar to the one shown.

	Temperature of Beaker		
	Cold	Medium	Hot
Initial Temperature			
Time of Diffusion			

Introduce the activity with a discussion of the importance of the movement of molecules into, out of, and within the cell. Define the term "operational definition" and explain that the students will form an operational definition of diffusion based on their observations. Their operational definitions can provide a basis for more detailed discussion later.

If hot plates are not available, water can be boiled in a teapot or coffee-maker. Water can be chilled over-night in a refrigerator, if the use of ice cubes is inconvenient. A 9°C spread of temperatures from hot to cold should produce easily observed differences in the rate of diffusion.

8. **NOTE:** Do not shake or stir the water. Add five drops of food coloring to the beaker. Note the time required for the coloring to spread uniformly through the water. Record the result.
9. Remove the ice cubes from the Cold beaker. If necessary, add cold water to the beaker until its level is about equal to that of other beakers.
10. Repeat steps 7 and 8 for the Cold and Medium beakers.
11. Clean up your materials and wash your hands before leaving the lab.

Analysis

1. *Defining Terms* An operational definition is a definition based on observation. State an operational definition of diffusion based on your observations of the food coloring in the beakers.

   Answers may vary. Diffusion should be defined in terms of the movement of molecules until they are uniformly distributed throughout the medium. Diffusion will always tend to produce uniform mixtures, since molecules move into areas where they are scarce.

2. *Analyzing Data* Explain how temperature affects diffusion.

   Answers may vary according to the data but should indicate that the rate of diffusion is proportional to the ambient temperature.

3. *Making Inferences* Why is a warm body advantageous for a living thing?

   Answers will vary. Accept any logical response that addresses the question. Insightful students should understand that warm body temperatures increase the rate of diffusion without any energy expenditure by the cell itself.

Name ______________________ Date ____________ Class ____________

FOCUS ACTIVITY

Chapter 5

Interpreting Labels: Stored Food Energy

In this activity, students compare the calorie contents of some common foods, using supplied data as well as the labels they have collected from other foods. Caloric values are standardized for 100-g samples and graphed to make comparison easier.

Background

Food supplies us with matter to build living tissue and energy to do work. The energy content of foods is measured in calories. How do foods differ in their energy content? In this activity you will interpret data from the labels of food products to answer this question.

Objectives

In this activity you will:

- ***graph*** and ***interpret*** data pertaining to the energy content of various foods

Materials

- food product labels (6)
- sheets of graph paper (2)

A few days before scheduling the activity, ask the students to form teams of two. Have each team collect the labels from at least six different food items.

Preparation

1. Form cooperative teams of two students. Complete steps 2–8.
2. Collect nutrition lists from six food products.
3. Make a table similar to the one shown below.

Food	Grams per Serving	Calories per Serving	Calories per Gram	Calories per 100 Grams

NOTE: You may substitute your choice of foods and calorie values.

Suggested Food List
Peanuts, 563 cal/100g
Ham, 95 cal/100g
Fudge cookies, 555 cal/100g
Crackers, 457 cal/100g
Corn, 71 cal/100g
Rice, 353 cal/100g

Procedure

Discuss the role of energy in living systems—ways in which humans use energy. Answer any questions students may have about the meaning of the terms. Some students may need assistance in converting ounces to grams and in making a bar graph.

Analysis

4. What types of nutritional information are included on your labels?

Answers will vary according to the foods sampled.

5. Select one label. Find the serving size and calories.
6. Calculate the number of calories per gram of food. Divide the serving size (in grams) by the number of calories per serving. If necessary, convert ounces to grams by multiplying the serving size by 28.4 g/oz.
7. Find the number of calories per 100-gram sample. Multiply calories per gram by 100. Record the results.
8. Repeat steps 5–7 for each of your labels.
9. Make a bar graph that compares the calorie content for 100-gram samples of each of the foods in your table.

1. **Analyzing Data** Which foods contain the most calories? The fewest calories?

Answers will vary according to the foods sampled. Of the foods initially listed in the table, peanuts and fudge cookies have the most calories, while corn and ham have the fewest calories.

2. **Evaluating Methods** Why is it helpful to convert the measurements to the same sample size?

Equal sample sizes make it easier to compare the calorie content for a given mass of food.

3. **Evaluating Methods** Why is calorie content on the label stated in terms of a serving size rather than in terms of a 100-gram sample?

The serving size on the label reflects the way most people should actually eat.

4. **Applying Concepts** How should a person put information about the calorie content to use?

People use nutritional information to plan a balanced diet.

Interpreting Information in a Pedigree

Schedule this activity prior to discussing the chapter or after a discussion of simple patterns of inheritance. The students construct and analyze pedigrees that summarize the family history of a trait or a set of traits in an easy-to-read diagram.

Background

Organizing information is often the key to solving a problem. Tracing the hereditary characteristics over many generations can be especially confusing unless the information is well organized. In this activity, you will learn how to organize hereditary information, making it much easier to analyze.

Objectives

In this activity you will:
• **construct** and **analyze** a pedigree

Materials

• paper
• pencil

Enhance this activity by providing examples of pedigrees. Medical genetics textbooks, the genetics department of a large hospital, and agricultural breeding services are some of the sources of such pedigrees.

Preparation

1. Pedigree I traces the dimples trait through three generations of a family. Blackened symbols represent people with dimples. Circles represent females and squares represent males.

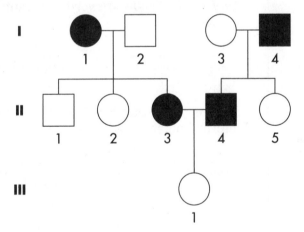

Pedigree I

Procedure

Introduce the activity by reading the hereditary information presented for one of the families in this activity. Have students describe the family without looking at their textbook. Their success should be minimal at best. Point out that the way information is organized can be crucial to using it and that this activity will show them a

2. The following passage describes the family shown in Pedigree I.

Although Jane and Joe Smith have dimples, their daughter, Clarissa, does not. Joe's dad has dimples, but his mother, and his sister, Grace, do not. Jane's dad, Mr. Renaldo, her brother, Jorge, and her sister, Emily, do not have dimples, but her mother does.

3. Write the name of each person below the correct symbol in Pedigree I. How are marriage and offspring symbolized? What do the Roman numerals symbolize?

A straight line connecting a circle and square indicates a marriage, with a descending
line leading to any offspring. Roman numerals identify each generation.

simple way to organize hereditary information.

Discuss the mechanics of Pedigree I before they continue with Pedigree II. Extend the activity by having the students construct pedigrees from summaries of novel situations, such as those involving maternal and fraternal twins, remarriage after divorce or death, and marriage across generations.

4. Make a pedigree based on the following passage about freckles.

Andy, Penny, and Delbert have freckles, but their mother, Mrs. Cummins, does not. Mrs. Giordano, Mrs. Cummins's sister, has freckles, but her parents, Mr. & Mrs. Lutz, do not. Deidra and Darlene Giordano are freckled, but their sister, Dixie, like her father, is not freckled.

Analysis

1. *Evaluating Techniques* What advantages does a pedigree have over a written passage?

Answers may vary but should indicate that a pedigree organizes hereditary information in less space, making it easier to read than the same information in a written passage.

2. *Summarizing Observations* How does a pedigree organize hereditary information, making it easier to understand?

Answers may vary but should indicate that a pedigree shows the number, gender, traits, and family relationship of each individual in a generation. All members of a generation are placed on the same line, identified by a Roman numeral. The symbols representing married individuals are joined by a straight line, with a perpendicular line descending to their offspring.

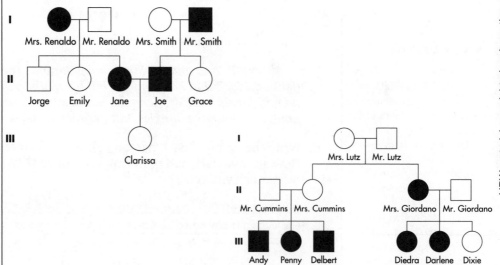

Making Models

In this activity, students learn to construct and manipulate models. If scheduled before reading or discussing the chapter, the activity can provide an opportunity for exploration rather than confirmation. The students are guided through the construction of one model that they analyze, and then they use it as a basis for completing a second model. After students have completed the activity, the model can provide the basis for a detailed discussion of the structure of DNA.

Background

You know that DNA provides instructions that direct the activities in a cell. In this activity, you will build a model of DNA to help you understand its structure and functions in the cell.

Objectives

In this activity you will:
• **construct** and **analyze** a model

Materials

• plastic soda straws (8)
• centimeter ruler
• scissors
• permanent marker

• pushpins (12 of each color), red, blue, yellow, and green
• paper clips (48)

Preparation

1. Form cooperative teams of four students. Select one member of your group to form a team. Work with your teammate to complete steps 2–9.
2. **CAUTION: Pointed objects can cause injury if not properly used.**
   Cut the soda straws into 3-cm segments to make 48 segments.

Procedure

Discuss how models help us recognize patterns and relationships.

Have students cut straws into 3-cm segments prior to beginning the activity. Small, standard-sized paper clips work well. Colored pushpins are available through office supply stores.

3. Insert a pushpin midway along the length of each straw segment. Push paper clip into one end of each straw segment until it touches the pin.
4. Keep the pins in a straight line, and insert the paper clip of a blue-pushpin segment into the open end of a red-pin segment. Add segments of straw to the red-segment end in the following order: green, yellow, blue, yellow, blue, blue, green, red, blue, and green. Use the permanent marker to label the blue segment on the end "top." This strand of segments is one-half of your first model.

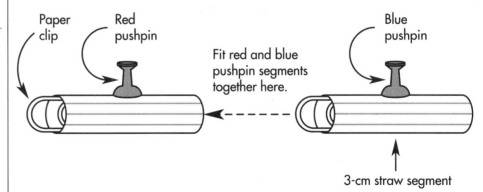

Paper clip   Red pushpin

Fit red and blue pushpin segments together here.

Blue pushpin

3-cm straw segment

5. Begin to construct the other half of your first model with a yellow-pin segment. Keep the pins in a straight line. Link segments together in the following order: green, red, blue, yellow, blue, yellow, yellow, red, green, yellow, and red. Label the yellow segment on the end "top."

6. Place the strands parallel to each other on the table with the "top" blue pin of one strand facing the "top" yellow pin of the second strand. What color pin is always across from a blue pin?

Yellow is always across from blue.

What color pin is always across from a red pin?

Green is always across from red.

7. Use 12 of the remaining straw segments with pins of any color, in any order, to make one-half of another model.

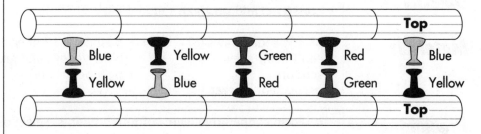

8. Exchange your team's 12 remaining straw segments and the strand that your team made in step 7 with those of the other team in your group.

9. Using the straw segments and the strand from the other team, make a strand of segments that has the correct sequence of color pins to complete your second model.

Analysis

1. **Comparing and Contrasting Models** How are your two models similar?

Answers may vary but should indicate that each model has pins of four different colors, such that blue faces yellow and red faces green.

How do they differ?

The models differ in how these colors are sequenced on a strand.

2. **Analyzing Models** What did you learn about the first model that enabled you to construct a second model?

Construction of the second model was made possible by recognizing that yellow is always paired with blue and that green is paired with red .

Making a Genetic Engineering Model

This activity gives students the opportunity to use models similar to those made in Focus Activity 7 to investigate processes relating to genetic engineering. This is meant to be an introductory activity that provides students with a concrete example of a sophisticated abstract process.

Background

In this activity, you will use simple models to demonstrate the manipulation of genetic material to produce new combinations of traits.

Objectives

In this activity you will:
- *construct* a model
- *demonstrate* processes related to genetic engineering

Materials

- soda straws, 3-cm pieces (50)
- centimeter ruler
- scissors
- pushpins (14 of each color), red, blue, yellow, and green
- paper clips (50)

Preparation

1. Form cooperative groups of four students. One 2-person team should complete steps 2 and 3 while the other team completes steps 4–6. Work with your entire group to complete steps 7–10.

Procedure

Small, standard-sized paper clips work well in this model. Colored pushpins are available through office supply stores. The students can cut straws into 3-cm segments at home or prior to beginning the activity.

Review the nature of the models to be used and the structure of DNA before beginning the activity. Do not attempt to introduce new vocabulary at this time.

2. **CAUTION: Pointed objects can cause injury if not properly used.** Make a model of a bacterial DNA molecule similar to the one described in Focus Activity 7. Arrange the nucleotides of the master strand in the following order: blue, red, green, yellow, red, red, blue, blue, green, red, blue, green, red, blue, blue, green, yellow, and red.

3. With your double-stranded DNA model lying on the table, form a circular molecule by carefully joining the opposite ends of each strand. Make a sketch of the molecule that shows the arrangement of the bases.

4. **CAUTION: Pointed objects can cause injury if not properly used.** Make one strand of a donor, human DNA molecule similar to the one described in Focus Activity 7. This strand should have the sequence: blue, blue, red, red, yellow, green, green, blue, red, and yellow.

5. Make a complementary strand of donor DNA having the sequence: blue, red, red, yellow, green, blue, yellow, yellow, green, and green.

6. Match the complementary portions of the two strands of your human DNA fragment. What is unusual about the structure of this donor DNA fragment?

Opposite strands at each end of the DNA fragment have four unpaired, exposed nucleotides.

Make a sketch of the donor molecule.

BBRRYGGBRY
   BRRYGBYYGG

7. Imagine that an enzyme moves around the circular molecule of bacterial DNA until it finds the sequence red-red-blue-blue and its complementary sequence green-green-yellow-yellow. Find this sequence in the bacterial molecule that you drew in Step 3.

8. Simulate the action of the enzyme by splitting the circular molecule at the sequence you identified in Step 7. Separate the yellow nucleotide from the blue at one end of the sequence, and the green from the red at the opposite end of the sequence on the complementary strand. Make a sketch of the split molecule.

9. Move the double-stranded donor, human DNA fragment into the break in the bacterial DNA molecule.

10. Imagine that a second enzyme joins the ends of the donor and recipient DNA creating a new DNA molecule. Make a sketch of the final bacterial DNA molecule.

Analysis

1. **Making Comparisons** How does the original bacterial DNA molecule differ from the final DNA molecule?

   The new bacterial molecule has more nucleotides than the original molecule as a result of the addition of the human DNA fragment.

2. **Applying Concepts** Of what possible benefit could this process be to humans?

   Answers may vary. A human gene could be inserted into a bacterial DNA molecule.

Comparing Observations of Body Parts

This activity gives the students the opportunity to make and analyze observations relating to homologous structures. Use it as the introduction to the chapter, or after the discussion of the evidence of evolution.

Background

Could you tell if two strangers were related just by looking at them? What kinds of evidence would help you determine their relationship? In this activity you will observe parts of various animals and look for evidence that these animals are related to one another.

Objectives

In this activity you will:
• *describe* structures of different organisms
• *identify* relationships between the structures of different organisms

Materials

• paper
• pencil

No additional materials are required. Skeletons of vertebrate limbs would enhance the activity.

Preparation

1. Make a table to record your observations of the limbs of seven different animals. Include columns across for the limb's *shape*, the *number of bones in the upper limb*, the *number of bones in the lower limb*, a description of the *arrangement of the bones*, and the limb's *function*. List the names of the animals down the left side of the table.

 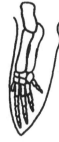 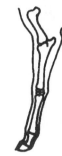 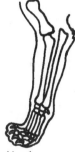 

Frog's foreleg   Whale's flipper   Horse's foreleg   Lion's foreleg   Penguin's flipper

Human's arm   Bat's wing   Bird's wing   Alligator's foreleg

Procedure

Stress that biologists often must rely on observations made from the drawings of others, at least during the initial stages of an investigation. Explain that in this activity, students must make the best use of the drawings of appendages.

Analysis

2. Record your observations in your table.

1. *Analyzing Observations* How are the limbs of the frog, whale, lion, human, bat, bird, and alligator similar?

Answers may vary but should indicate that the number and arrangement of bones in the upper and lower limbs of the vertebrate animals are similar.

How do they differ?

The shape and function may differ.

2. *Analyzing Observations* Which limbs perform similar functions?

Appendages of the bird and bat are used for flight. The human limb is used for grasping and manipulation of objects, the whale's and penguin's in swimming, and the others' in locomotion on land.

3. *Identifying Relationships* Which is the better indicator of the relationship between two organisms—structure or function? Explain your reasoning.

Answers may vary but should suggest that structure is a better indicator of relationship than is function.

Analyzing Adaptations: Living on Land

This activity gives students an opportunity to observe some of the adaptations that allowed life to move out of the water. As students compare the structures of various pairs of aquatic and terrestrial organisms, they should observe the need to avoid desiccation and the need to acquire oxygen—important factors in moving to a terrestrial environment.

Background

The move from a watery environment to the land was a giant step in evolution. In this activity you will study some of the adaptations that allowed organisms to make the move to land.

Objectives

In this activity you will:
• **observe** various structural features
• **relate** structural features to the move to a terrestrial environment

Materials

• various aquatic and terrestrial specimens

Preparation

Place pairs of specimens around the laboratory for inspection. Each pair should consist of one terrestrial specimen and one aquatic specimen showing an adaptation for life on land. Include the name of each organism and any additional directions needed to draw your students' attention to the related adaptation.

1. Form cooperative teams of two students. Complete steps 2–4.
2. Make a table similar to the one shown below for recording your observations.

Specimen Pairs		Observations	Adaptions for Life on Land
Name	Habitat		
1.	Aquatic		
	Terrestrial		
2.	Aquatic		
	Terrestrial		
3.	Aquatic		
	Terrestrial		
4.	Aquatic		
	Terrestrial		

Procedure

Discuss the differences the students perceive between terrestrial and aquatic environments.

3. Pairs of specimens are displayed around the classroom. One member of each pair is identified as an organism that lives in the water while the other lives on land. Observe one pair of specimens. Record their names and your observations.
4. Review your observations and suggest reasons why one organism is better adapted for life on land than the other.

Analysis

1. **Summarizing Data** List the features of land organisms that make them better adapted for life on land.

Answers may vary according to the specimens observed but might include roots, a waxy covering on plants, a waterproof exoskeleton, a seed with a seed coat and stored food, and lungs.

2. *Analyzing Data* Based on your observations, describe two factors that organisms had to overcome to survive out of water.

Accept any logical answer. Desiccation and the need to obtain oxygen are the two most

important obstacles to overcome in order to survive on land.

NOTE: A wide variety of specimens are available from WARD'S

Pairs of specimens could include any or all of the following:
• a chicken's egg and a frog's egg to compare shelled to nonshelled eggs
• an insect and a flatworm, jellyfish, or sponge to show the cuticle or the exoskeleton of the insect
• a seed and a spore to show the seed coat and food supply
• organisms with lungs and those that use diffusion or gills to obtain oxygen
• a bryophyte and a plant with an extensive root system
Another pair of specimens can include a fresh apple and an apple that has been dipped briefly in gasoline to remove the cutin. Allow both apples to sit for two days before this activity. The apple without the cutin should be shriveled.

Name _____ Date _____ Class _____

Comparing Primate Features

Structural similarity is one indicator of evolutionary relationship. In this activity, students compare the skull, jaw, and spinal attachment, and hand of the gorilla with those of the human. The similarities should suggest the existence of a common ancestor.

Background

Paleontologists are scientists who look for and study fossils. Structural similarities can indicate an evolutionary relationship between organisms. In this activity you will observe structural features of primate hands, jaws, and skulls.

Objectives

In this activity you will:
• **observe** primate structures
• **relate** these features to evolutionary relationships

Materials

• paper
• pencil

No materials are required.

Preparation

1. Form cooperative teams of two students. Complete steps 2–7.
2. Make a table similar to the one shown below for recording your observations.

	Comparison of Human and Ape Features	
	Human	Gorilla
Hand		
Jaw		
Skull		
Spinal attachment		

Procedure

Review the sources of evidence of evolutionary relationship presented earlier.

3. Compare the structure of the gorilla's hand below to your own hand. Try to touch the tip of your thumb to the tip of every other finger on that hand. Does it appear that a gorilla could do this?

Human hand

Gorilla hand

4. Record your observations of the similarities between these hands.

5. Compare the jaws of the human and the gorilla. Describe the similarities and differences.

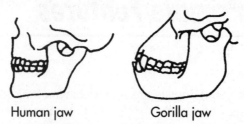

Human jaw   Gorilla jaw

6. Compare the structures of the human skull and the gorilla skull. Describe the similarities and differences.

Human Skull   Gorilla Skull

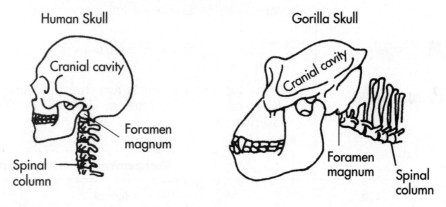

7. Use the figure above to compare how the spinal column attaches to the skulls of the human and the gorilla. Describe the similarities and differences you observe.

Analysis

1. **Summarizing Observations** Summarize the evidence suggesting that apes and humans may have had a common ancestor.

   Answers may vary. Students should describe such similarities as the opposable thumb, the number and shapes of teeth, and the general shape and structure of the skull. Differences include the shape and size of the jaw, the attachment point of the spinal column to the skull, the size of certain teeth, and the number of teeth.

2. **Making Inferences** How does the structure of the hand contribute to the ability to use tools?

   An opposable thumb allows primates to grasp tools more tightly.

3. **Identifying Relationships** How does skull structure provide evidence of a primate's relative intelligence?

   Answers may vary. Students should suggest that the size of the portion of the cranium that surrounds the brain indicates the size of the brain. This factor, in turn, may provide evidence for inferences about an organism's relative intelligence.

Making a Food Web

This activity introduces students to the concepts of food chains and food webs. The activity draws on everyday experience and common sense rather than on detailed knowledge of a particular ecosystem.

Background

One organism consumes another for energy and raw materials. A food chain shows the sequence in which energy passes from one organism to another as it moves through the community.

Objectives

In this activity you will:
- *categorize* organisms
- *make* a food web

Materials

- paper
- pencil

No materials are required. Slides, transparencies, or photographs showing organisms common to a woodland habitat can be helpful.

Preparation

1. Form cooperative groups of two students to complete this activity.
2. Closely observe the illustration on this page showing a portion of a community. List all the organisms that you see.

A woodland community

3. Add to your list other organisms that might also be present in this community, but are not shown.

Answers may vary. Students should list the organisms shown in the illustration and
additional organisms likely to be present in a woodland habitat.

Procedure

Discuss the types of organisms that might inhabit the community and be depicted in the activity. Feel free to substitute a local ecosystem for the one shown or to extend the activity by using an ecosystem more familiar to your students.

Analysis

4. On a separate sheet of paper, write the name of one organism from your list that is capable of photosynthesis.
5. Draw a short arrow leading from this organism to the name of a second organism that might eat it.
6. You have drawn the first two links of a food chain. Extend your chain to three links by adding an arrow and a third organism that might consume the second.
7. Extend your food chain to five links.
8. Make two more food chains consisting of five links each.
9. Construct a food web by adding arrows that connect organisms in different chains. Make as many connections as possible.

1. **Comparing Data** How are food chains similar to food webs, yet different?

   A food web is composed of individual food chains. It includes links between food chains that in themselves form additional food chains.

2. **Evaluating Methods** Why can a food web be more helpful than a description of the same information written in a paragraph?

   A food web presents the data in a graphical representation, making it easier to see relationships among the various links.

Using Random Sampling

In this activity students compare data obtained from an estimate by sampling with data obtained from an actual count. Emphasize that estimating by sampling can yield reasonably accurate population data. Have the students conduct an actual field study or come up with techniques to reduce the percentage error in their estimates.

Background

Scientists cannot possibly count every organism in a population. One way to estimate the size of a population is to collect data by taking random samples. In this activity you will look at how data obtained by random sampling compare with data obtained by an actual count. Sampling is used to track population growth in an ecosystem. It is one of many methods used by scientists to collect data when studying ecosystems.

Objectives

In this activity you will:
- *estimate* by sampling
- *count* the number of sunflower plants in a meadow
- *compare* gathered data
- *calculate* percentage error

Materials

- unlined paper
- ruler
- pencil
- containers (2)
- scissors

Preparation

Any small containers, such as bowls or large culture dishes, can be used to hold the paper, which can be reused from class to class.

1. Cut a sheet of paper into 20 slips, each approximately 4 cm x 4 cm.
2. Number 10 of the slips from 1 to 10, respectively. Put the numbered slips in a small container.
3. Label the remaining 10 slips from A through J and put them in a second container.
4. To record your data, make a table similar to the one shown.

Method	Average number of plants per grid segment	Total number of plants
Sample		
Actual count		

Procedure

For those flowers that fall on the grid lines, have students include plants that fall on the upper line and right-side line of each square. Ignore those that fall on the lower and left side line.

5. The grid shown here represents a meadow measuring 10 m on each side. Each grid segment is 1 m × 1 m. Each black circle represents one sunflower plant.

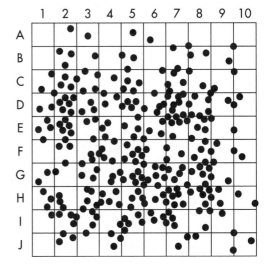

6. Randomly remove one slip from each container. On a separate sheet of paper, write down the number-letter combination you drew and find the grid segment that matches that combination. Count the number of sunflower plants in that combination. Count the number of sunflower plants in that grid segment. Record this number on the separate sheet of paper. Return each slip to the appropriate container.

7. Repeat step 6 until you have data for 10 different grid segments. These 10 grid segments represent a sample. Gathering data from a randomly selected sample of a larger area or group is called *sampling*.

8. Find the total number of sunflower plants for the 10-segment sample. This is an estimate, based on sampling, of the number of plants in the meadow. Divide this number by 10 to determine the average number of sunflower plants per square meter in the sample. Record this number in the table. Multiply the average number of sunflower plants per square meter by 100 to find the total number of plants in the meadow. Record this number in your table.

9. Count all the sunflower plants actually shown in the meadow. Record this number in your table. Divide this figure by 100 to calculate the average number of sunflower plants per square meter. Record your data.

Analysis

1. **Comparing Data** Compare the actual count with the data you recorded for your random sampling estimate.

   Answers will vary according to the data collected. There are 261 sunflower plants
   actually represented in the meadow.

2. **Analyzing Methods** Why was the paper-slip method used to select grid segments?

   Answers may vary but should suggest that the paper-slip method is a way to select the
   segments randomly.

3. **Evaluating Methods** What is the percentage error in your estimate? To calculate percentage error find the difference between the actual count and the estimated count. Divide this difference by the actual count and multiply by 100.

   $$\frac{\text{Difference}}{\text{Actual count}} \times 100 = \% \text{ Error}$$

   Answers may vary, but low percentage error should suggest that estimating by sampling can provide data that are reasonably close to actual counts.

4. **Making Inferences** How could you change the procedure in this activity to reduce your percentage error?

   Answers may vary, but students should recognize that making sure that the sample is truly
   random and not localized in any way reduces percentage error. Also, increasing
   the number of random samples provides data that are closer to the actual counts, thereby
   reducing the percentage error.

Determining The Amount of Refuse

This activity gives students firsthand knowledge of recycling and the problems associated with generated refuse.

Background

How much refuse does a family produce every day? How much of this material could be recycled? In this activity you will collect data to answer these questions.

Objectives

In this activity you will:
- *infer*, *measure*, and *compare* the amount of refuse your family produces each day

Materials

- disposable gloves
- apron
- goggles

- plastic garbage bags
- bathroom scale or other weighing device (at home)

Preparation

Each student should have access to a bathroom scale for measuring the mass of refuse produced. If this is not possible, give the students small plastic refuse bags. Have the students measure in "refuse bag" units. Afterward, you can help them convert to a standard volume unit or to mass using a conversion factor based on the average mass in a typical bag of refuse.

1. How many kilograms of refuse do you think your family produces each day? (NOTE: Each pound is equal to 0.454 kg.)
2. Make a table similar to the one shown below.

	Type of Refuse		
Day	Recyclable	Reusable	Waste
1			
2			
3			
4			
5			
6			
7			
Average			
Total (yr)			

Procedure

3. **CAUTION: Wear disposable gloves, an apron and goggles while completing the following activity.** Use fresh refuse from within your own home only; do not work with refuse more than one day old or with refuse from outside your home.
4. Sort through your waste and garbage containers at home. Separate the contents into three groups: materials that could be recycled, materials that could be reused, and materials that are truly waste.
5. List the items in each group.
6. Place the items in each group in a plastic garbage bag.

7. Weigh each bag on a bathroom scale. Multiply the weight in pounds by 0.454 to obtain the weight in kilograms. Record your data. NOTE: If you do not have the use of a bathroom scale, see your teacher for special instructions.
8. Repeat steps 3–7 each day for 7 days.
9. Calculate the average for each group of materials and record your answer in your table.
10. Calculate the total amount of yearly refuse produced by your family by multiplying the sum of the averages of the three groups of refuse by 365. Record your results in your table.
11. Clean up your materials and wash your hands before leaving the lab.

Analysis

1. **Summarizing Data** Summarize your data.

Answers will vary.

2. **Analyzing Data** How would recycling affect the total amount of refuse produced by your family?

Answers will vary but should indicate that recycling would significantly reduce the amount of refuse produced.

3. **Making Inferences** Why would re-use and recycling programs benefit our society?

Answers may vary but should indicate that these programs should benefit all of society by placing less demand on resources and the environment, as well as adding to the economic base.

Have students prepare a list naming the types of things that can be recycled. Remind them that nonaluminum cans and plastic bottles can also be recycled. You might also have them locate the local recycling facility.

Direct the students to collect their data each day from the refuse within their own home. Caution them not to work with refuse that is more than one day old because of health hazards.

Name ______________________ Date ____________ Class ____________

FOCUS ACTIVITY

Chapter 15

Grouping Things You Use Daily

The process of classification is introduced in this activity by having students consider a system for grouping their own possessions. Classification systems are encountered every day; food stores, department stores, recorded-music stores, and warehouses are just a few of the places that commonly employ a system of classification.

Background

Classification systems are part of our everyday lives. You find things placed in groups in food stores, in record stores, and even in your home. In this activity, you will explore how you group things you use daily.

Objectives

In this activity you will:
- ***identify*** types of things that you group
- ***explain*** why you group things

Materials

Each team of students will need:
- paper
- pencil

Preparation

Although no materials are required for this activity, visual aids, such as a table, suggest that the use of classification systems in everyday life will increase the relevance of this activity to the student.

1. Make a table similar to the one shown below.

Level		
I	II	III
A. Clothing	1.	a.
		b.
	2.	a.
		b.
	3.	a.
		b.
	4.	a.
		b.
B. School	1.	a.
		b.
	2.	a.
		b.
	3.	a.
		b.
	4.	a.
		b.
C. Recreation	1.	a.
		b.
	2.	a.
		b.
	3.	a.
		b.
	4.	a.
		b.

2. Close your eyes and picture your possessions. Think of all the things that you use in everyday life.

3. List your possessions. Be as specific as possible.

Procedure

Emphasize the process of classification and its application in everyday life. Some students might find it easier to make their initial list of possessions while at home. Students can share their characteristics, but avoid having them share their lists of possessions.

4. Divide your list into three smaller lists—clothing, school-related items, and recreational items.
5. Divide each list into four subgroups. Use logical characteristics as the basis for your subgroups. For example, casual clothing or books might be two subgroups.
6. Record the characteristics used for these subgroups in the Level II column of your table.
7. List the items that belong to each subgroup.
8. Where possible, divide each subgroup into two groups. Use logical characteristics as a basis for these groups and record this information in Level III of your table.
9. List the items that belong to each of these groups.

Analysis

1. **Analyzing Observations** Do the characteristics used to distinguish the various groups become more general or more specific as you move from Level I to Level III?

   They become more specific.

2. **Evaluating Techniques** Explain how you might use a grouping or classification system of your possessions.

   Answers will vary but should suggest that systems are often grouped to make them easier to find.

3. **Applying Concepts** Explain how classification systems are used by merchants or other businesses.

   Answers may vary but should suggest that merchants group items so shoppers can find them more easily.

Using Bacteria to Make Food

While students and the general public are very much aware of the harmful effects of bacterial contamination, they are less familiar with the beneficial uses of microorganisms. In this activity, students make a simple yogurt culture. With reasonable attention to cleanliness and good lab technique, the yogurt can be tasted after its production.

Background

Although bacteria can cause disease and destroy other forms of life, they also have many beneficial uses. In this activity, you will have the opportunity to use bacteria to make a food product.

Objectives

In this activity you will:
- **observe** the production of yogurt

Review the concept of pH and acids and bases. Demonstrate the use of pH test strips if the students are unfamiliar with them.

Materials

Each team of students will need:
- apron
- disposable gloves
- milk, 400 mL
- beaker, 600-mL
- pH test strips
- hot plate

- beaker, 1,000-mL
- water
- thermometer
- yogurt (plain, with active cultures)
- teaspoon
- aluminum foil

Preparation

1. Form cooperative teams of four students. Complete steps 2–8.
2. What is pH and how is it measured?

Procedure

Commercial yogurt can be used as a starter culture. Active yogurt cultures commonly contain *Streptococcus thermophiles* and *Lactobacillus bulgaricus*.

Only clean glassware should be used. The inoculated milk cultures should be incubated at about 39°C.

3. **CAUTION: Put on a laboratory apron and disposable gloves.** Leave them on throughout this activity.
4. To avoid contamination, keep your work area and equipment as clean as possible. Clean all glassware, spoons, and thermometers thoroughly before beginning this activity and upon its completion.
5. Pour 400 mL of milk into a clean 600-mL beaker.
6. Measure the pH of the milk with a pH test strip. Record the results.
7. **CAUTION: The hot plate produces high temperatures that can cause a serious burn.** Place the beaker of milk in a 1,000-mL water bath. Heat the milk to 81°C for 15 minutes. Be careful to avoid boiling the milk.
8. Allow the milk to cool to about 39°C.
9. Add one teaspoon of yogurt to the milk and stir gently until mixed. Cover the beaker with aluminum foil.
10. Incubate the milk at about 39°C for 24 hours.
11. Measure the pH of the newly formed yogurt. Record your results.

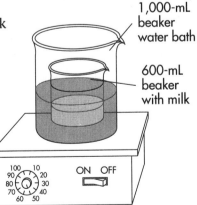

1,000-mL beaker water bath

600-mL beaker with milk

pH is the concentration of the hydronium ions of a solution, expressed as the negative of the common logarithm. The pH of a solution is measured on a pH scale that ranges from 0 to 14, measuring the acidity or alkalinity of a solution.

Analysis

1. **Summarizing Observations** Describe the changes that occurred in the milk.

   Answers may vary but should describe a change in consistency and a change in pH in the culture. The milk's initial pH of about 7.0 will decrease as the bacterial action makes the culture slightly acidic.

2. **Analyzing Data** What evidence do you have that a chemical reaction occurred in the milk?

   Answers may vary but might describe the change in consistency and the change in pH as evidence that a chemical reaction has occurred.

3. **Applying Concepts** What other evidence could be collected to indicate that yogurt was produced in the beaker?

   Answers may vary but might suggest that a microscopic comparison of the new culture and the starter culture could provide evidence that yogurt was produced. Also, a taste and smell test could provide additional evidence.

Observing Protists

This investigation gives students an opportunity to observe cultures of identified protists. Discuss techniques that can improve observations. Have students explore why drawing and taking notes can improve their powers of observation. Discuss the benefits of listing items to look for, and the benefits of searching for relationships among observations.

Background

A glass of water taken from a pond can contain thousands of microorganisms. In this activity you will have the opportunity to view some examples of these organisms.

Objectives

In this activity you will:
• *observe* and *compare* characteristics among protists

Materials

Each team of students will need:
• medicine droppers
• methyl cellulose
• coverslips
• yeast-Congo red solution
• *Paramecium* culture
• glass slides
• compound light
• additional protist cultures

Review the prelab questions, the proper use of the microscope, and the preparation of a wet mount, if necessary. Remind students that adjusting the diaphragm can darken the field, making unstained organisms easier to observe.

Preparation

Obtain a variety of protists from WARD'S. *Paramecium, Blepharisma, Amoeba, Euglena, Stentor, Vorti-cella,* and *Volvox* are just a few suitable genera. Select genera that exhibit a variety of structural characteristics, methods of locomotion, and types of feeding responses.

1. Name the five kingdoms and give an example of each.

_____

_____

_____

_____

How can you distinguish between the organisms of each kingdom?

_____

_____

_____

_____

2. Make a table similar to the one shown below to record your observations of protists.

Protist Name	Sketch	General Description	Type of Movement	Feeding Mechanism

3. Review the procedure for making a wet mount.
4. Form cooperative teams of two. Complete steps 6–8.

Procedure

Prepare the yeast-Congo red solution by adding one package of dry yeast to 100 mL of water. Let the mixture stand for 24 hours. Then bring the mixture to a boil and add 0.2 g of Congo red powder. Boil this mixture for about 8 minutes, being careful not to burn it. Allow the solution to cool before using it.

Analysis

5. Place a drop of methyl cellulose on a clean glass slide to slow the movement of the protists. Place a drop of *Paramecium* culture on the methyl cellulose. Gently add a coverslip.
6. Use low power to locate the organisms on the slide. Switch to high power and focus on one *Paramecium* for several minutes. Record your observations in your table.
7. In your table, make a labeled drawing of this organism. Include the organism's name and the magnification that the drawing represents. Write a brief description of your observations next to your drawing.
8. Using a clean dropper for each culture, repeat steps 4 through 7 using each of the other cultures. Record your observations in your table.
9. Clean up your materials and wash your hands before leaving the lab.

1. ***Analyzing Observations*** What characteristics are common to all of the organisms that you observed?

Answers may vary but should indicate that all the specimens were unicellular and possessed a nucleus bounded by a nuclear membrane.

2. ***Analyzing Observations*** In what ways do these organisms differ?

Answers will vary according to the specimens selected but should discuss such characteristics as size, color, feeding response, and method of locomotion.

3. ***Analyzing Observations*** What evidence do you have that some protists might be capable of photosynthesis?

Some specimens are green, suggesting the presence of chlorophyll.

4. ***Inferring Conclusions*** How might a protist's ability to move relate to its method of obtaining nutrition?

Answers should exhibit an understanding that locomotion allows the specimen to search for food. Protists that are sessile must wait for food to approach or be capable of making their own food.

5. ***Identifying Relationships*** Based on observable characteristics, make a simple biological key that could be used to identify the organisms you observed.

Answers will vary, but students should be able to use such characteristics as color, method of locomotion, and feeding response to develop a key.

Comparing Plant Adaptations

This activity gives students the opportunity to observe a variety of plant adaptations. Emphasize observation rather than naming plant structures. Based on their observations, students are asked to draw inferences about the evolutionary advantages of these adaptations.

Background

Plants are found in almost every habitat on Earth. They exhibit a variety of sizes, shapes, and structures. In this activity you will observe a variety of plants.

Objectives

In this activity you will:
• **observe** and **compare** vascular and nonvascular plants

Materials

Each team of students will need:
• A number of plant specimens

Preparation

1. Form cooperative teams of two students. Complete steps 3–10.
2. Use a separate sheet of paper to prepare the following table for recording your observations. Across the top, on the first line, center and write the heading: *Characteristics.* On the second line write the following headings: *Size in cm; Multicellular Structure; Habitat; Roots, Stems,* or *Leaves; Flowers; Seeds; Evidence of Internal Tubes; Waxy Covering on Leaves; Openings in Surface of Leaf;* and *Division of Labor.* Down the left side, write the number and name of the specimen.

Procedure

Discuss the nature of the traits to be observed in the lab and the evidence that would indicate the existence of each trait in a particular specimen. Review any procedures to be used that are unique to the specific specimens being used.

3. Observe all the demonstration materials at your lab station. These may include a living plant, a microscopic preparation, or graphic materials.
4. Based on your observations of the materials, record information about size, multicellular structure, habitat, and the presence of roots, stems, leaves, flowers, and seeds in your table.
5. Observe the surface of the living specimen and the surface of the microscopic preparation, if one is available, for evidence of internal tubes. Record your observations.
6. Observe the surfaces of the specimens, if possible, for evidence of a waxy covering. Record your observations.
7. If available, observe the microscopic preparation of a leaf. Look for evidence of openings in the surface. Record your observations.
8. Is there evidence of a division of labor in this plant? Record your answer in your table.
9. Repeat steps 3–8 for each of the plants assigned by your teacher.

Analysis

1. **Summarizing Observations** Summarize your observations of the plants.

Answers will vary according to the specimens observed.

_____

_____

2. **Identifying Relationships** What traits seem to be common to all plants that have roots, stems, or leaves?

Evidence of internal tubes, multicellularity, and division of labor are common to all plants with roots, stems, and leaves.

_____

3. ***Making Inferences*** What traits appear to be common to all plants living in a dry habitat?

Evidence of a waxy covering, seeds, internal tubes, multicellularity, and division of labor are common to plants from a dry habitat.

4. ***Analyzing Data*** Describe the traits of a highly evolved plant and explain your reasoning.

Answers will vary but should describe all the traits observed in this activity as being present in a highly evolved plant.

Make a variety of specimens available for observation. Include live specimens, microscopic preparations, and graphic materials as necessary. Provide clues to direct the students' attention to structural features you feel they may otherwise miss. Include specimens of algae, mosses, liverworts, ferns, gymnosperms, and angiosperms.

Inferring Function From Structure

This activity gives students the opportunity to explore the structure of a leaf and to make simple inferences based on their observations. Develop interest in the chapter rather than introducing the details or terminology of leaf anatomy.

Background

A leaf is a complex structure composed of a variety of tissues working together to help the plant survive. The primary function of a leaf is food production. In this activity, you will explore the structure of a leaf and make inferences about leaf function based on your observations.

Objectives

In this activity you will:
• *infer* function from *observations* of leaf structure

Materials

Each team of students will need:
• lilac leaf cross section, prepared slide
• microscope

Prepared slides of lilac leaf cross sections can be obtained from WARD'S (91 M 8720).

Preparation

1. Form cooperative teams of two students. Complete steps 2–7.
2. List the functions of vascular tissue in plants.

   Vascular tissue in plants transports water, food, and minerals throughout the plant.

3. List the specific tasks a leaf must perform to carry out its primary function.

   Specific tasks should correspond with photosynthesis.

Procedure

Review the process of photosynthesis and the general structure of a vascular plant. Have students list what they know of the functions performed by the roots, stems, and leaves.

4. View a prepared slide of a lilac leaf cross section under low power. Find an area of the leaf that looks similar to the drawing below. If necessary, move the slide to observe the entire length of the section.
5. Observe this section of the leaf under high power.
6. Make a drawing showing the structure of this section of the lilac leaf.
7. Describe this section of the leaf in detail. Pay particular attention to the areas labeled A–F below, but base your description on your specimen.

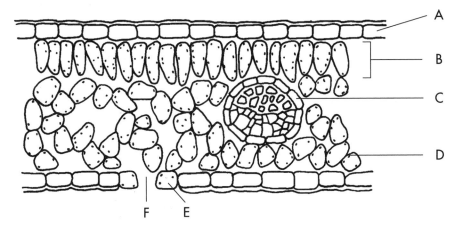

Lilac leaf cross section

Analysis

1. *Using Relationships* How do the location and structure of area A in the leaf provide evidence of its function?

Answers may vary but should suggest that area A covers the inner tissues and provides

protection.

2. *Making Inferences* What does the shape of structure C suggest about its possible function?

The round, tubelike shape suggests that area C functions as tubes to transport materials to

and from the leaf.

3. *Making Inferences* What do the contents of the cells in area B suggest about their function?

The cells in area B contain green structures, suggesting that photosynthesis occurs in this

area. Also this area's location just beneath the upper surface provides the most light, as

well as protection from the outer environment.

4. *Identifying Relationships* How might carbon dioxide enter the leaf? Explain your reasoning.

Answers may vary but should suggest that carbon dioxide could enter through opening F

of the drawing, in the bottom of the leaf.

Relating Root Structure to Function

This activity gives students the opportunity to investigate the relationship between the root structure and the soil-anchoring ability of two food plants. The loss of soil to water and wind erosion depletes the amount of land available for growing food products. Plants reduce the amount of soil lost to erosion.

Background

Erosion depletes the amount of land available for growing food products. Plants that provide food while binding the soil are of special value to us. In this activity, you will investigate the relationship between the root's structure and its ability to hold the soil.

Objectives

In this activity you will:
• **analyze** data
• **evaluate** inferences

Materials

Each team of students will need:
• paper
• lima bean seedlings (3)
• corn seedlings (3)
• adhesive tape
• scissors
• cord
• hand-held spring scale
• centimeter ruler

Preparation

1. Form cooperative groups of four. Work with one member of your group to complete steps 3–11. One team in the group will use lima bean seedlings while the other team uses corn seedlings.
2. On a separate sheet of paper make a table similar to the one shown below.

Seedling No.	Lima Bean Seedlings			Corn Seedlings		
	Force g	Length mm	Diameter mm	Force g	Length mm	Diameter mm
1						
2						
3						
Average						

Seeds and pots are available from garden supply stores and WARD'S (Beans: 86 M 8003, Corn: 86 M 8080, Pots: 20 M 2130, Soil: 20 M 8306). Soaking seeds in a 1:4 solution of bleach and water for 15 minutes will reduce loss due to bacterial and fungal contamination. Have the class plant the seeds in advance of the scheduled activity. Seedlings may take three to four weeks to reach a height of 10 cm. Use potting soil to reduce the breaking of stems as the plant is pulled from the soil. A hand-held spring scale is available from WARD'S (15 M 3775).

Procedure

Demonstrate how to read the scale. Caution students to pull the plants with a steady motion. A snapping motion is more likely to break the stem and would be very difficult to measure.

3. Select three pots of the same type of seedling. Place a small piece of adhesive tape around each stem at the level of the soil.
4. Attach one end of a cord to the tape and the other end to a spring scale.
5. Carefully, while reading the indicator, pull the scale upward and slightly away from the plant, as shown in the drawing below.
6. As the plant is uprooted, note the force on the indicator. Record the results.
7. Place the plant on a piece of paper.
8. Measure the length and diameter of the root system. Record your results.
9. Repeat steps 4–8 for each plant.
10. Find the average force, length, and diameter for your sample.
11. Combine your data with that of the other team in your group.
12. Use the averages to make a bar graph that compares the measurements for both kinds of plant.

Spring scale

Plant

Cord

Tape

Pot

Analysis

1. ***Summarizing Data*** Summarize the data in your graphs.

   Answers may vary but should adequately summarize the data.

2. ***Identifying Relationships*** Describe any evidence of a relationship between the force required to pull the plant from the soil and the other measurements.

   Answers will vary according to the data collected. There will likely be a relationship
   between the force required to pull the plant from the soil and the length and diameter of
   the root system.

3. ***Making Predictions*** Which root system would be most effective in an area where wind erosion could be a problem? Explain your answer.

   Plants such as corn and other grasses have fibrous roots that bind soil effectively, making
   them suitable for areas where erosion can be a problem.

4. ***Identifying Relationships*** Why is knowledge of plant structure helpful to a farmer when choosing a crop's location?

   Answers may vary but should suggest that a knowledge of plant structure would help a
   farmer make wise decisions about the location of crops.

Recognizing Patterns of Symmetry

This activity introduces students to the concept of symmetry. They observe an assortment of animals and group them based on body plan. More formal definitions of radial and bilateral symmetry can be developed later. At this stage, develop the concept of symmetry.

Background

Symmetry is the regular arrangement of parts around a point on an object or body. In other words, symmetry can be thought of as a plan for the arrangement of body parts. In this activity you will try to recognize the types of symmetry found in the animal kingdom.

Objectives

In this activity you will:
• **observe** animals
• **identify** different types of symmetry

Materials

Each team of students will need:
• index cards (one per specimen), 3 x 5 in.
• an assortment of animals

Preparation
Procedure

1. Form cooperative teams of two students. Complete steps 2 through 6.

2. Closely observe one animal.
3. Write the animal's name on an index card.
4. Draw a simple sketch of the animal showing its external appearance.
5. Write a brief description of the animal below its sketch.
6. Repeat steps 2–5 for each animal you observe.
7. Sort your cards into three groups—A, B, and C—based on the symmetry of the animals.

Analysis

Discuss the definition of symmetry presented in the activity. Have students give examples of symmetry or lack of symmetry in objects they encounter in everyday life. Do not introduce the terms radial and bilateral at this time.

1. **Summarizing Observations** Describe the traits common to each of your groups of animals.

   Answers will vary depending on the specimens used but should relate to the body plans
   observed.

2. **Identifying Relationships** The word "radial" means extending outward from a center point. Why can this word be used to describe one type of symmetry in animals?

   Answers will vary but should explain how animals such as the sea star, jellyfish, and sea
   urchin have parts that radiate from a central point on the body.

Have students move from station to station to observe an assortment of animals showing asymmetry, and radial and bilateral symmetries. Specimens might include the following: a number of different sponges, a jellyfish, a sea urchin, a sea star, a snail, a grasshopper, a crayfish, a frog, a lizard, a gerbil, or other choices. Use live specimens whenever possible. Have students supply their own index cards. A variety of specimen may be purchased from WARD'S.

3. *Identifying Relationships* The word "bilateral" means having two sides or halves. Why can this word be used to describe one type of symmetry in animals?

Answers will vary but should explain how animals such as the crayfish, grasshopper, and

gerbil have parts that form two sides that appear to be mirror images of each other.

4. *Making Inferences* Sponges are generally considered to be asymmetrical. What must this description mean?

Answers may vary but should suggest that the word "asymmetry" means the lack of an

organized body plan.

Comparing Animal Eggs

In this activity, students compare the egg of an animal that develops on land to the egg of an animal that develops in the water. The relative proportion of food stored in the egg and the existence of a hard shell are two adaptations that should be easily observed.

Background

To survive in the harsh environment of the land, animals evolved adaptations relating to reproduction and development. Some of these adaptations can be observed in the structure of their eggs.

Objectives

In this activity you will:
• **compare** and **contrast** eggs of land and water animals

Materials

Each team of students will need:
• small spoon
• preserved frog eggs
• culture dish
• water
• hand lens or stereomicroscope
• blunt probe
• forceps
• chicken eggs

Preparation

Review the requirements for maintaining life. Caution students to treat the frog eggs gently.

1. Form cooperative teams of two students. Complete steps 2–16. Use a separate sheet of paper to make a table for recording your observations and drawings.
2. What does an embryo require to develop normally?

   food, water, protection, a means of gas exchange, and waste removal

   _____

3. What functions does an egg perform?

   The egg provides a suitable environment for the developing embryo until it is able to exist

   independently.

   _____

Procedure

4. Use a small spoon to transfer a frog egg to a clean culture dish. Add enough water to cover the egg.
5. Observe the egg using a hand lens or the highest magnification on your stereomicroscope. Record your observations.
6. Use a probe to gently touch the egg's surface. Record your observations.
7. Make a drawing of the frog egg. Show the relative proportion of the egg's light and dark material.
8. In general, the darker portion of the egg will develop into the embryo. The lighter portion of the egg will provide food for the developing cells. What fraction of the egg provides food for these living cells?

   Answers will vary depending upon the amount of the lighter portion which decreases

   during embryonic development.

   _____

Frog eggs (available only March–April) can be obtained from WARD'S (87 M 8205). Order three to four eggs per team, to allow for damage. The eggs can be reused, but expect some damage. Chicken eggs can be obtained from a grocery store. Fertile eggs are not required for this activity.

9. Return the frog egg to its storage container.
10. Observe the exterior of a chicken egg. Record your observations.
11. Add tap water to a clean culture dish until it is about one-quarter full. Gently crack the eggshell on the edge of the dish and separate the shell, allowing the contents to flow into the water.

12. Examine the rounded end of the eggshell's exterior and interior surfaces. Record your observations.

13. Use your forceps to examine the small white disk on the yolk's surface. This structure contains the living egg cell that would develop into the embryo. The remainder of the egg provides food and water for these cells. What fraction of the egg provides food and water for the living cells?

Answers will vary. Students should suggest the majority of the egg.

14. Draw the contents of the egg as it appears in the dish.

15. Write a detailed description of the contents of the egg.

16. Clean up your materials and wash your hands before leaving the lab.

Analysis

1. *Making Comparisons* Compare and contrast the frog and chicken eggs.

Students should discuss the absence or presence of a shell, relative size, and relative proportion of food in each egg.

2. *Identifying Relationships* Describe the structure of the eggshell and explain its purpose.

Students should discuss the presence of a thin membrane along the inner surface and the hard exterior covering of the eggshell, which protect the embryo from foreign material and drying. The students should suggest that since the frog develops in water, a shell is not required.

Why is a shell necessary for the survival of the chicken embryo but not for the frog?

Oxygen must be able to move into the egg, while carbon dioxide must move out.

3. *Identifying Relationships* What factors might explain why a chicken egg contains so much more food and water than the frog egg?

Answers may vary but should suggest that more food may be required for the chicken because it is larger and its time of development is much longer than that of the frog.

Observing Some Major Animal Groups

In this activity the students are asked to observe a small group of diverse specimens based on a list of traits of their own choosing. The focus of this activity should be on their choice of traits rather than on the specimens themselves.

Background

Since science is based on observation, knowing what to observe is very important to your eventual success. In this activity, you will observe a variety of animals and identify the traits you believe to be most meaningful.

Objectives

In this activity you will:
• *observe* some major groups of animals
• *evaluate* traits in terms of their importance in describing animals

Materials

Each team of students will need:
• one specimen each of a number of different kinds of animals

Preparation

Each team of two students should have a minimum of six specimens to observe. While the students can move from station to station observing single specimens, having all six at one station will facilitate comparisons. The specimens used by each team can vary, although a diverse group is recommended. Try to include one specimen from each phylum. Refer to WARD'S catalog for a variety of specimens.

The Preparation questions point the students in the direction of two important traits used to compare animal groups: highest level of organization and symmetry. Review these concepts before continuing with the activity. Allow students to select their own list of traits to observe so they can discover that some traits may be more meaningful than others.

1. Form cooperative groups of four students. Work with one member of your group to complete steps 2–6.
2. List and define the levels of organization exhibited by living things, from the lowest to the highest.
3. Describe and give an example of each type of symmetry exhibited by animals.
4. Make a table similar to the one shown below. **NOTE:** Your headings may differ from those on the sample table below.

Trait	Animal Specimen					
	Sponge	Jellyfish	Flatworm	Earthworm	Crayfish	Frog
Highest Level of Organization						
Symmetry						

Procedure

5. Observe the animal specimens.
6. List and describe six traits that you can use to describe these animals. For example, each animal can be described in terms of the highest level of organization it exhibits.
7. Share your list of traits with the other team in your group. Discuss the traits and decide on the best four to use to describe animals. Record these four traits in the first column of your table.
8. Observe your assigned animals in terms of the traits in your table. Record your observations.

Analysis

1. *Summarizing Observations* Describe, in general terms, how these animals are similar and how they differ.

Answers will vary but should discuss the traits of the six specimens.

2. *Evaluating Methods* Which trait or traits would you say are most important?

Answers will vary but should suggest that some traits may have a bearing on others.

Explain your reasons.

For example, the level of organization is more significant than the existence of a specific organ system.

3. *Evaluating Methods* Which trait or traits would you say are less important in describing animals?

Accept any logical answer.

Explain your reasons.

4. *Applying Concepts* Based on your experiences, what factors are important in describing animals?

Answers will vary but should be logical. Traits that affect the existence of other traits are the most significant.

Explain your reasons.

These traits are generally of greater evolutionary significance because they have such a major bearing on the adaptive advantage of the organism.

Observing Insect Behavior

This activity gives students the opportunity to observe simple adaptive behaviors in crickets. Use this activity to stimulate interest for further study of insect structure and behavior.

Background

With more than 750,000 species, insects may be the most successful group of animals on Earth. This activity allows you to observe some aspects of structure and behavior that help make insects so successful.

Objectives

In this activity you will:
• **observe** cricket behavior
• **locate** and **identify** common external structures

Materials

Each team of students will need:
• cricket
• beakers (2), 500 mL
• plastic wrap
• apple
• masking tape
• aluminum foil
• self-locking freezer bags (2)
• crushed ice
• hot tap water

Preparation

1. Form cooperative groups of four students. Work with one member of your group to complete steps 2–12. Use a separate sheet of paper for recording your observations.

Procedure

2. **CAUTION: Use humane treatment when handling live specimen.** Place a cricket into a clean, 500-mL beaker and quickly cover the beaker with plastic wrap. The supply of oxygen in the beaker is sufficient for you to complete your work.
3. Add a small piece of apple to the beaker. Set the beaker on the table. Sit quietly for several minutes and observe the cricket. Any movement will cause the cricket to stop what it is doing. Record your observations.
4. Remove the plastic wrap from the beaker and quickly attach a second beaker. Join the two beakers together, at the mouths, with masking tape.
5. Wrap one beaker with aluminum foil.
6. Gently tap the sides of the beaker until the cricket is uncovered. Lay the joined beakers on their sides with a bright lamp over the uncovered beaker.
7. Without disturbing the cricket, carefully move the aluminum foil to the other beaker. Observe the location of the cricket after five minutes. Record your observations.
8. Fill a plastic freezer bag half full with crushed ice. Fill another bag half full with hot tap water. Seal each bag and arrange them side by side on the table.

Caution the students to use care and to be gentle while handling the crickets. Discuss the appropriate methods for collecting and transporting the animals while in the lab.

Live crickets can be purchased from WARD'S (87 M 6100) or, in many cases, from local pet stores and bait shops. Store crickets in cricket cages or in a number of small aquariums. Line the bottom of the aquarium with coarse sand or sawdust. Place a shallow pan containing water and a sponge on the floor of the aquarium. Small pieces of potato, apple, or dry dog food will provide nourishment.

9. Gently rock the joined beakers from side to side until the cricket is in the center. Place the joined beakers on the freezer bags as shown below.

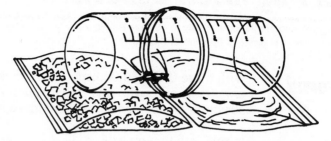

10. Observe the cricket's behavior for five minutes. Then record your observations.
11. Perform steps 9 and 10 two more times. Then record your observations.
12. Compare your observations with those of the other team in your group. Discuss the possible reasons for any differences.
13. Clean up your materials and wash your hands before leaving the lab.

Analysis

1. **Making Inferences** Explain why giving crickets small dishes filled with sugar solution would not keep them alive.

   Answers will vary but should suggest that the cricket's mouth is designed for chewing rather than lapping or sucking.

2. **Identifying Relationships** What evidence suggests that crickets are sensitive to light?

   Answers may vary but should describe the movement of the cricket from areas of bright light to darkness.

3. **Making Inferences** How does the cricket's response to temperature improve its chance of survival?

   Answers may vary but should relate the cricket's need for adequate warmth to its response to temperature.

Observing a Frog

In this activity the students make simple observations of the external anatomy of a common amphibian—the frog. While most students have seen frogs, many have not had an opportunity to study a live frog closely. Approach the activity as an exploration that can provide a concrete basis for further detailed discussion.

Background

Amphibians can be found in a variety of habitats. You have probably seen a live frog or toad before, but perhaps you have never closely observed one. In this activity, you will observe the features of a common amphibian—the frog.

Objectives

In this activity you will:
• **observe** amphibians
• **describe** characteristics

Materials

Each team of students will need:
• large plastic tub
• paper towels
• beaker, 250 mL
• water
• frog
• beaker, 1,000 mL
• gauze
• cotton swab

Preparation

1. Based only on past experience, describe in detail the structure and general appearance of a frog. Make a table for recording your observations and drawings.
2. Form cooperative teams of three. Complete steps 3–13.

Procedure

Caution the students about the proper handling of the specimens. The frogs need to be kept moist to prevent injury. Ask students to close their eyes and picture a frog before they actually view the specimens. Have them write a description of the image they see. Emphasize that this activity focuses on observations; they should not hurry, thinking that they have already seen all there is to see.

3. Cover the bottom of the tub with three layers of paper towels. Fill the 250-mL beaker with tap water for wetting the towels. (**NOTE:** Keep water on hand to keep towels from drying.)
4. **CAUTION: Use humane treatment when handling live specimen.** Use a 1,000-mL beaker to carry a frog to your lab table.
5. Place the frog on the wet towels in the center of the tub. Cover the frog with the beaker. Wait 3–5 minutes while the frog calms. Avoid moving or touching the glass.
6. Slowly remove the beaker and observe the frog. Make a detailed drawing showing its external features. Label the head, abdomen, and appendages.
7. Observe the frog's head. Look closely at its eyes and look for evidence of ears. Record your observations by adding to your drawing.
8. Watch the frog as it breathes. Record your observations.
9. Look closely at the frog's forelimbs and hind limbs. Observe its fingers, feet, and toes. Record your observations in additional drawings.
10. Run your fingertip along the frog's skin. Compare the way it feels with the way your skin feels. Record your observations.
11. Gently prod the frog with the tip of a moistened cotton swab. Observe its movement. Pay special attention to the way the individual parts of its legs bend. Show your observations in a sketch of the frog's movement.

Live frogs can be obtained from WARD'S (87 M 8217) and some local live bait stores, or you may have experienced students collect some samples. Ten frogs should be enough for a typical class. If the frogs are to be used by many classes, have enough specimens to allow a period of rest between exposures to the students.

12. Return the frog to the beaker and then to the aquarium.
13. Clean up your materials and wash your hands before leaving the lab.

Analysis

1. **Summarizing Observations** Summarize your observations by writing a detailed description of a frog's anatomy.

Answers will vary but should provide a more detailed description of the frog than that presented in the preparation.

2. **Making Comparisons** How is a frog similar to you?

Similarities might include symmetry, the number of eyes, and the movement and number of appendages.

How is a frog different from you?

Answers will vary. Students could discuss differences associated with the attachment of the head to the body, in the number of fingers, and the parts of the leg.

Vertebrate Skeletons

This activity gives students the opportunity to observe vertebrate skeletons and to make inferences about evolutionary relationships. Skeletal features provide easily observable evidence of homologies that form the basis for these inferences.

Background

Skeletal features can reveal relationships among animals. Similarities provide evidence that different animals evolved from a common ancestor. Bones also provide evidence of ecological relationships that exist among animals. In this activity, you will observe a number of vertebrate skeletons and make inferences based on your observations.

Objectives

In this activity you will:
• **compare** vertebrate skeletons
• **infer relationships** based on homologies

Materials

Each team of students will need:
• a variety of skeletal specimens or other resources

Preparation

While the activity can be completed using photographs and drawings of skeletons, actual specimens (a perch, a pigeon, a garter snake, a turtle, a cat, a frog, and a human skeleton) are recommended. One specimen of each can be shared by the entire class.

1. Form cooperative teams of two students. Complete steps 2–10. Use a separate sheet of paper for recording your observations.

2. What are homologous structures?

   A homologous structure is a body part with the same basic structure as that of another

   organism, suggesting common ancestry.

3. Describe the type of evidence that might suggest that your hand, a pigeon's wing, and a cat's paw are homologous.

   Answers may vary but should suggest that the number, arrangement, and general shape of

   the bones that make up the human hand, pigeon wing, and cat paw are sources of
   evidence of homology.

Procedure

Have students move around the lab stations, recording their observations of various specimens. They can complete the activity at their desks, returning to a specimen if necessary.

4. Examine the available models and illustrations of vertebrate skeletons below.

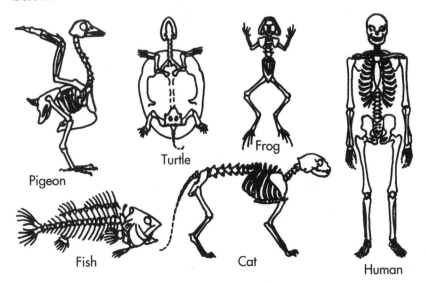

Pigeon

Turtle

Frog

Fish

Cat

Human

5. List four similarities and three differences that you observe.
6. Examine the backbone of each skeleton and describe their similarities and their differences.
7. Which two animals have backbones that are most alike?

With exception of the tail, the cat and the human.

8. A human upper limb includes the shoulder, upper arm, forearm, wrist, and hand. How many bones make up each segment?

Three bones make up the shoulder, one bone the upper arm, two bones the forearm, eight bones the wrist, and twenty bones the hand.

9. Observe the bones of the wrist and hand. Record your observations.
10. Compare the human upper limb with the forelimbs of the cat, the pigeon, the turtle, and the frog. Record your observations.

Analysis

1. **Analyzing Observations** Which animal's forelimb is most like the human upper limb?

The human upper limb and cat forelimb are most alike.

Describe the observations that led you to this conclusion.

The bones are similar in number, arrangement, and relative size.

2. **Analyzing Observations** What evidence do you have to support the inference that vertebrates evolved from a common ancestor?

Much homology is observed in the skeletons of vertebrates. Many similarities in the numbers, arrangements, and relative sizes of the bones of various structures support the hypothesis that vertebrates evolved from a common ancestor.

3. **Analyzing Observations** Based on your observations in this lab, why should humans and cats be considered more closely related than humans and any of the other vertebrates you observed?

Human skeletons and cat skeletons should demonstrate the greatest degree of homology.

How is the classification of humans and cats consistent with this conclusion?

Students should suggest that both cats and humans are mammals sharing many characteristics.

Comparing Skeletal Joints

Students explore five types of skeletal joints in this activity. They observe and analyze the movements of the major types of skeletal joints. Next, they conduct a survey of other joints in the body, identifying each according to its type.

Background

The bones of your skeleton come together in *joints*. In this activity, you will explore the types of skeletal joints found in your body by observing an example of each type of joint and surveying the other joints in your body.

Objectives

In this activity you will:
• *identify* and **compare** skeletal joints

Materials

Each team of students will need:
• paper
• pen
• human skull or other animal skull

Have a real or artificial skull to illustrate immovable or fixed joints. The skull of any animal can be substituted for a human skull. A detailed illustration will suffice, if necessary.

Preparation

1. Use a separate sheet of paper to make a table for recording your observations of the skeletal joints shown below.

Pivot joint

Hinge joint

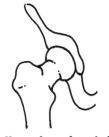

Ball-and-socket joint

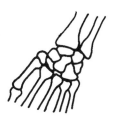

Gliding joint

Fixed cranial joint

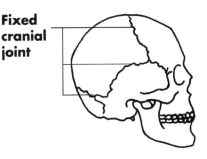

Procedure

Tell the students to use common sense and care while making the movements described in the activity. Only normal movements are required. Review the procedure for examining the movement of the pivot joint. Holding the neck is necessary to illustrate the limited side-to-side movement of this joint.

Analysis

2. **CAUTION: Do not move in ways that can cause injury or pain.** Straighten your index finger as if to point. Slowly bend the finger until it touches your palm. Moving only your finger, try to move the individual parts of the finger in other directions. Record your description of the movements in your table.
3. Move your arm in as many ways as possible. Record your description of the movements first from the shoulder, then from the elbow.
4. Place your hands on the sides of your neck to hold the neck in place. *Gently* move only your head in all possible directions. Try not to let your neck bend. Record your observations of the movements.
5. Use your left hand to grip your right arm just above the wrist. Without moving your forearm, move your hand in all possible directions. Record your observations of the movements.
6. Examine the joints in the top of a skull. Gently press on each side of the joints to see if movement is possible without damaging the skull. Record your observations.
7. Starting with your feet, examine the movement of each of the other joints in your skeleton. Record the other locations in the body where you discover each type of joint.

1. **Comparing Observations** Rank the five types of joints according to their freedom of movement. Start with the joint that allows the least freedom of movement.

   Fixed cranial joint, pivot joint, glide joint, hinge joint, and ball-and-socket joint

2. **Analyzing Observations** Which types of joints are involved in walking?

   Ball-and-socket joint, hinge joint, and glide joints

3. **Applying Concepts** For each type of skeletal joint, name a common, nonliving object that has a similar type of joint in its construction.

   Answers might include the expansion joint in bridges, a beacon, small disc or ball
   attached to the underside of furniture legs to allow easy sliding, door hinges, ball-and-socket wrenches, and lamps.

Procedure Answers

2. Hinge joints allow the index finger to move in a forward and backward direction.
3. A ball-and-socket joint allows the wide movement between the arm and shoulder, whereas the elbow is a hinge joint with limited movement.
4. Rotating movement by a pivot joint allows the movement between the head and the neck.
5. Gliding joints allow movement of the wrist.
6. Fixed cranial joints allow no movement.

Name _____ Date _____ Class _____

Bias and Experimentation

Touch relies on two types of receptors, one type located deep in the dermis, the other located just below the epidermis. This activity allows students to determine the distribution of these touch receptors, using a simple procedure that reduces the amount of error caused by the expectations of the subject.

Background

A biased answer is based on your beliefs rather than on observation. The possibility of bias exists when a person's evaluation is required in an experiment. For example, you see an object touching the skin but do not actually feel it. Because you expect to feel it, you might actually say that you feel the object. Your answer is biased by your expectations. In this activity, you will try to reduce the effect of such bias in determining the distribution of touch receptors in the skin.

Objectives

In this activity you will:
• **measure** the distance between touch receptors
• **understand** the need to **reduce bias** in experimentation

Materials

Each team of students will need:
• cardboard
• scissors
• dissecting pins (11)
• metric ruler

Preparation

Use clean dissecting pins and caution the students to only gently touch the pins to the skin's surface. Thick corrugated cardboard holds the pins upright, keeping their tips the measured distance apart.

1. Form cooperative teams of two students.
2. Cut cardboard into six 3×5 cm rectangles.
3. Make a *touch-tester* by inserting two pins, 2 mm apart, halfway through one piece of cardboard, as shown below.

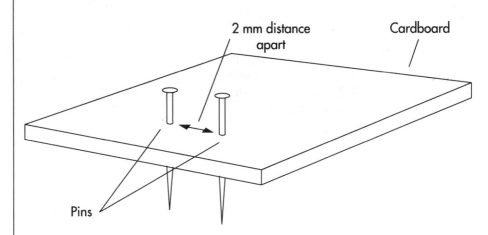

2 mm distance apart   Cardboard

Pins

4. Place the other pairs of pins 5 mm, 1 cm, 2 cm, and 3 cm apart in the other pieces of cardboard. Push the last pin through the center of the last rectangle. Each touch-tester is used to touch the tips of the pins to the skin.
5. When two pins are so close together that they stimulate only one touch receptor, a person feels only one pin. *How could these touch-testers be used to find the distance between touch receptors?*

   Answers will vary but should suggest a method in which the minimum distance is found
   when two objects touched to the skin can be sensed as two objects.

6. Record your data in a table similar to the table below.

Distance Between Pins	Number of Pins Perceived When Touched					
	Back of Neck	Back of Hand	Fingertip	Inside Forearm	Outside Forearm	Lips
2 mm						
5 mm						
1 cm						
2 cm						
3 cm						

Procedure

Discuss the procedure for determining the distribution of touch receptors.

CAUTION: Test subject must close eyes and remain motionless. Tester must apply the pins gently. Avoid piercing the skin.

7. Use the touch-tester to gently touch both pins to the back of your partner's neck. Record a "+" if two pins are perceived and a "-" if one pin is felt. Repeat the test for three trials.

8. *What is the minimum distance at which the pins can be consistently perceived as two objects?*

Answers will vary but should be in the range of 5 mm to 1 cm.

Later in the activity, have the class discuss techniques that could be employed to reduce this type of error. Explain how double-blind techniques are used to reduce the effect of bias.

9. Since the subject expects to feel two pins, a possible source of error may arise. The subject's response may be biased. *How could using the rectangle with the single pin reduce this source of error?*

Answers should suggest that a single pin should be occasionally used so that the subject does not know what to expect.

10. Using all six touch-testers, repeat steps 7 through 9 for each area of the body indicated in the table.

11. Exchange roles and repeat the tests.

12. Use your data to make a bar graph showing the average distance between touch receptors.

Analysis

1. *Summarizing Data* Summarize your data.

Answers will vary but should describe a variation in distances between touch receptors for different areas of the body.

2. *Analyzing Data* Account for differences between the data obtained for you and for your partner.

There is likely to be some variation in data between subjects, but the pattern of distances between body areas is likely to be maintained.

3. *Applying Methods* Describe a situation where expectations could produce erroneous data.

Answers will vary but should describe a situation when the expectations of the subject lead to an erroneous response. For example, a patient might think that a medication relieved the symptoms of an illness, even though only a placebo was administered.

Graphing Growth Rate Data

Students compare data for height versus age in human males and females by making a graph. An analysis of the data reveals a similar pattern of growth for both sexes before and after puberty begins, which suggests that sex hormones may play a role in the process.

Background

Have you ever glanced at the family photo album and considered the developmental changes your body has undergone? If so, you have probably wondered what biological event triggered these transformations. Chemical messengers within your body dictated these changes. In this activity, you will analyze the relationship between age and human growth.

Objectives

In this activity you will:
• *graph* and *analyze* human growth rate data

Materials

Each team of students will need:
• graph paper
• blue pencil
• red pencil

Preparation

Review the procedure for making a line graph.

1. Using graph paper, prepare a grid for graphing the data in the table shown below.

Age	Average Height in Centimeters	
	Males	Females
5	109	109
7	119	119
9	133	133
11	138	145
13	157	160
15	169	162
17	177	162
19	177	163

Procedure

2. Use a blue pencil to make a line graph that compares height to age for human males.
3. On the same graph, use a red pencil to make a line graph that compares height to age for human females.

Analysis

1. *Analyzing Data* Use the data to explain whether humans grow at a continuous rate, or in spurts.

Suggest that growth does occur in spurts especially around the age of puberty.

_____

2. **Analyzing Data** Explain how the growth pattern for males differs from that of females.

The data indicate that growth rates for males and females are similar until puberty. The growth rate of females increases prior to that of males, as does the onset of puberty.

3. **Making Inferences** What evidence suggests that sex hormones might play a role in growth patterns?

The increase in growth rate at the time of puberty for each gender is evidence that sex hormones may play a role in growth.

4. **Evaluating Methods** How does the graph make interpreting the data easier?

The relationship between age and growth rate for both males and females is easier to observe in the graph than in the table of data.

Collecting Data Through a Survey

This activity fosters group cooperation by having students conduct a survey relating to health and the use of tobacco, alcohol, and other drugs. This emphasizes student interaction and cooperation rather than concepts specifically relating to health.

Background

The advancement of science depends on the interaction and cooperation of human beings. This activity requires you to work together in a cooperative effort to collect data through a survey of your peers.

Objectives

In this activity you will:
• work together to **design** a survey and **analyze** data

Materials

Each team of students will need:
• paper
• pencil

Preparation

1. Make a list of six statements that might be used in a survey to test a person's knowledge about health and the use of tobacco, alcohol, and other drugs. Three statements should be true and three statements should be false.
2. Form cooperative teams of four. Complete Steps 3–8.

Procedure

Students may write the statements during class or as a homework assignment. Encourage the students to consider relevant statements that truly test a person's understanding of important aspects of the topic. It is important that students research the statements. Emphasize the importance of writing well-researched statements to ensure that the survey is valid.

3. List your team's six statements on a sheet of paper.
4. Use your text or library references to check that half of the statements are true, and that half of the statements are false.
5. Your team should choose 12 statements for use in your survey. Have each member of the team make a survey data sheet similar to the model below.

Group: _____			
Statement	**True**	**False**	**Total**
1.			
2.			

6. Choose two different groups of people to interview. For example, you might compare students of different age groups, smokers and non-smokers, or adults and adolescents.
7. Have each member of the team interview three people from each group. Have each interviewee identify whether they believe each statement in the survey is true or false. Record each response in the appropriate column by making a check (✓).
8. Pool your team's data. For group I, record the total number of responses and the number of correct responses to each statement. Then calculate the percentage of correct responses for each statement.
9. Repeat step 8 for group II.

10. Make a bar graph similar to the sample shown below to compare the percentages of correct responses to each statement made by each group.

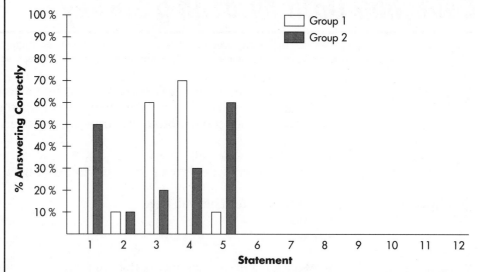

Analysis

1. **Summarizing Data** Summarize the data collected through your survey.

   Answers should adequately summarize the data collected through the survey.

2. **Evaluating Methods** List and describe the benefits of working in a group to complete this activity.

   Answers will vary but should discuss benefits such as the advantages of sharing in ideas and labor.

3. **Evaluating Methods** List and describe any problems that must be overcome when working in a group.

   Answers should discuss such problems as the difficulty in reaching a consensus and dealing with diverse personalities.

4. **Analyzing Data** Did you find any difference between the overall knowledge of the two groups?

   Answers will vary according to the data collected.

5. **Making Inferences** If any difference exists between the two groups, how would you account for it?

   Answers will vary according to the data collected.

6. **Analyzing Data** On which statement(s) did the subjects show the greatest need for information?

   Answers will vary according to the data collected.

Name _____ Date _____ Class _____

Determining Lung Capacity

In this activity the students use a balloon to measure vital capacity, expiratory reserve, and tidal volume. Volume is calculated based on the diameter of the balloon (a sphere). While some error is expected using this method, the results will give a good estimation. Have students participate in this activity on a voluntary basis. Individuals with respiratory problems should not be forced to participate.

Background

Health workers accurately measure lung capacity using an instrument called a spirometer. These measurements provide one source of information about the general health of the lungs. In this activity, you will measure your lung capacity using a balloon. While this method is less accurate than a spirometer, it does provide a good indication of your lung capacity.

Objectives

In this activity you will:
• **measure** lung capacity

Materials

Each team of students will need:
• round balloon
• metric ruler
• graph paper
• pencil

Preparation

Only round balloons should be used. For health reasons, have a fresh balloon for each student.

Review the technique for measuring the balloon's diameter. Remind students that the measurement of tidal volume should involve normal breathing only. Discuss the calculations involved in this activity. Some students may need help in making these calculations.

1. Form cooperative groups of four students. Work individually to complete steps 2–11.
2. Make a table similar to the one shown below for recording your data.

Trial	Balloon Diameter Measurements		
	Vital Capacity	Expiratory Reserve	Tidal Volume
1			
2			
3			
4			
Total (all trials)			
Average Diameter			
Volume cm^3			

Procedure

3. Stretch the balloon to make it easier to fill.
4. Take a number of deep breaths. Exhale as much air as possible into the balloon. Hold the end of the balloon shut to stop the air from escaping.
5. Lay the balloon on the table next to the ruler. Measure its diameter in centimeters and record your data as *Vital Capacity* for Trial 1. Release the air from the balloon.
6. Repeat steps 4 and 5 to complete Trials 2 through 4.
7. Breathe normally a number of times. Exhale normally. Then before inhaling again, force air in your lungs into the balloon. Measure and record the balloon's diameter as your *Expiratory Reserve*. Repeat three times.
8. Breathe normally a number of times. Take a normal breath and exhale a normal amount of air into the balloon. Measure the balloon's diameter as your *Tidal Volume*. Repeat three times.
9. Calculate the average diameter of the balloon for each group of measurements. Record the results.

10. Use the following formula to calculate the volume of air in the balloon.

    Volume $= 1.33 \pi r^3$,

    where r $= \frac{1}{2}$ of the average diameter of the balloon,

    and $\pi = 3.14$

11. Share your data with the other members of your group. Use all the data to make a bar graph comparing the three average lung volume measurements for each person in your group.

Analysis

1. **Summarizing Data** Summarize your data.

   Answers will vary according to the individual.

2. **Defining Terms Operationally** Use your work to define vital capacity, expiratory reserve, and tidal volume.

   Vital capacity is the maximum amount of air that is held in the lungs. The expiratory reserve is the amount of air that remains in the lungs after a normal exhalation. The tidal volume is the amount of air exchanged during a normal breath.

3. **Evaluating Methods** Describe any sources of error related to your final measurements of lung volume.

   Answers will vary but could include sources of error relating to the measurement of the balloon, error associated with overcoming the resistance of the balloon, and errors in the calculations.

4. **Analyzing Data** How do your measurements compare with those of the other members of your group?

   Answers will vary but should indicate that body size, gender, age, physical condition, and the use of tobacco can contribute to variability between individuals.

   Explain the reasons for any differences.

An approximate measure of "normal" vital capacity based on height, age, and gender can be found using either of the following formulas.

Vital Capacity$_{MALE}$ in cm^3 $= 52H - 22A - 3600$
Vital Capacity$_{FEMAL}$ in cm^3 $= 41H - 18A - 2690$
where:
H = height in centimeters (without shoes)
A = age in years

Differences of 10 percent or less are not considered significant.
Have students compare the calculated vital capacity with their measured vital capacity.

Name _____ Date _____ Class _____

Relating Cell Structure to Function

Blood is composed of cells, cell fragments, and fluid plasma. Students use the microscope to examine some of the components of human blood. Based on these observations, they relate blood cell structure to function.

Background

Blood is composed of cells, cell fragments, and fluid plasma. In this activity, you will relate the structure of specific blood cells to the function each performs.

Objectives

In this activity you will:
- *observe* blood cells
- *relate* cell structure to function

Prepared slides of human blood can be obtained from WARD'S. Alternatively, a 35-mm slide of a blood smear can be projected for the entire class to use. Having the students prepare slides of their own blood is not recommended.

Materials

Each team of students will need:
- prepared slide of human blood
- compound light microscope

Preparation

1. Form cooperative teams of two students to complete this activity. Use a separate sheet of paper to make a table for recording your observations.

Procedure

Review the basic structure of a cell. Draw the students' attention to the functions of the nucleus, cytoplasm, and cell membrane. Prepared slides may be obtained from WARD'S (93 M 6541).

2. Examine a prepared slide of human blood under low power of the microscope. Change the position of the slide and continue your examination.
3. Switch to high power and focus clearly on the cells. The most numerous cells are called red blood cells. Make a detailed drawing that shows a number of red cells. Label the parts of the cell that you recognize. Record your observations of the size, shape, and the relative number of the red cells.
4. Slowly move the slide until you find one or more larger cells that appear very different from the red cells. These cells are called white blood cells. Make a drawing of the white blood cell. Label the parts of the cell that you recognize. Record your observations of the size, shape, and relative number of white blood cells.
5. Look for evidence of tiny, dotlike cell fragments called platelets. Make a drawing of a few platelets. Record your observations of the size, shape, and relative number of platelets.

Analysis

1. *Identifying Relationships* The transport of oxygen throughout the body is one important function of blood. What evidence suggests that this function would be better performed by red blood cells than by white blood cells?

The enormous number of red blood cells, compared with white blood cells, suggests that they function in the transport of oxygen throughout the body. The absence of a nucleus reduces the mass of the cell, while the concave shape increases the surface area.

2. *Making Inferences* What evidence suggests that only white blood cells are capable of reproducing?

The absence of a nucleus in red blood cells suggests that they are incapable of reproduction.

3. *Identifying Relationships* The capture and destruction of foreign organisms, such as bacteria, is another function of blood. What evidence suggests that white blood cells are better suited to this function than red blood cells?

The large size of white cells suggests that they are better suited for engulfing foreign particles. Also, the presence of a nucleus suggests that white cells have the ability to reproduce, which would be important in fighting invading organisms.

Reading Labels: Nutritional Information

In this activity, students analyze nutritional information collected from food labels. Data for vitamin, mineral, carbohydrate, fat, protein, and caloric content are collected and organized in a table. This activity can be scheduled to introduce the unit or can be done after nutrition has been discussed.

Background

By law, the labels on food packages must list important nutritional information. Understanding and using this information can improve your health and can lead to better values in your food purchases.

Objectives

In this activity you will:
• **analyze** the information listed on the labels of food products

Materials

Each team of students will need:
• 6 labels from assorted food products

Preparation

About one week before this activity is to be done, have students collect labels from a variety of food products. The amount of nutritional information recorded on a label varies from product to product. Extra labels should also be on hand in the classroom.

1. Collect labels from six food products. Be sure to include the portion of the label that lists nutritional information.
2. Make a table similar to the one shown below.

Food Product	Vitamins	Minerals	Carbo-hydrates	Proteins	Fats	Calories

3. Form cooperative groups of four students. Work with one member of your group to complete steps 4 through 8.

Procedure

Review with the class the nutritional information found on food labels.

4. Place the name of a food product in the first column of your table.
5. Use the "Nutrition Information" on the label to find the vitamin content of the food. List each vitamin present in the food in your table. Record the percentage of the recommended daily allowance (RDA) for each vitamin that you listed.
6. Record the mineral contents of the foods in your table.
7. Record the carbohydrate, protein, and fat contents.
8. Record the calorie content.
9. Repeat steps 4–8 for each of the food product labels collected by your group.

Analysis

1. **Analyzing Data** In which foods do you find the greatest variety of vitamins?

   Answers will vary according to the data collected.

   _____

   _____

2. *Analyzing Data* In which foods do you find the greatest variety of minerals?

Answers will vary according to the data collected.

3. *Making Inferences* Which foods are the most nutritious?

Answers will vary according to the data collected but should indicate foods that are higher in vitamins, minerals, and protein while being lower in calories and fat.

Name _____ Date _____ Class _____

Culturing Frog Embryos

In this activity, students culture frog embryos and evaluate the effect of water temperature on embryo development. The exact temperatures used in this investigation are arbitrary, although temperatures below 4°C and above 28°C are likely to be lethal. To achieve a range of temperatures, embryo cultures may be placed in the refrigerator, in an incubator, and on a shelf in the classroom.

Background

Organisms in the early stages of development are called embryos. Scientists have learned that vertebrates share common genetic instructions for embryo development and are affected in a similar manner by changes within their environment. In this activity, you will culture frog embryos to study the effects of temperature on embryonic development.

Objectives

In this activity you will:
• **evaluate** methods for culturing frog embryos

Materials

Each team of students will need:
• beakers (3), 100-mL
• pond water
• hot water
• frog embryos
• stereomicroscope
• watch glass
• culture dish
• wax pencil
• spoon
• thermometer °C

Preparation

Frog eggs can be obtained from WARD'S (87 M 8205), available in March–April. Have all other materials ready at the time the embryos arrive. Aquarium water or aged tap water can be substituted for pond water.

1. Describe how water temperature might affect the development of frog embryos.

2. Make a chart similar to the one shown below for recording data.
3. **CAUTION: Put on a laboratory apron and disposable gloves.** Leave them on throughout this activity.

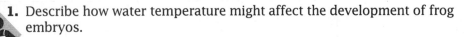

| Temperature | Observations of Tadpole Development | | | | |
| | Day | | | | |
	1	2	3	4	7
Room temperature					
Warm					
Cold					

4. **CAUTION: The hot plate produces high temperatures that can cause a serious burn.** Use the hot plate to warm the water. Use beakers to store warm, cold, and room-temperature water. Label each beaker accordingly.

Procedure

5. **CAUTION: Use humane treatment when handling live specimen.** Place frog embryos in a clean watch glass and cover them with pond water.
6. Use a stereomicroscope to observe the embryos. Notice the embryos' coloration. Record your observations. Make a drawing of your specimens.

Observing the ongoing development of any embryo can be a formidable task. Discuss strategies for making a problem more manageable. The first step in managing a complex problem is to divide it into a number of smaller, simpler tasks. Observing and describing such specific features as size, shape, and apparent cell number should seem more manageable.

7. Remove specimens that are severely injured or dead. *Why will removing these embryos benefit the remaining specimens?*

Answers should suggest that the dead or dying specimens are likely to increase the bacterial and fungal contamination of the water.

8. Label a culture dish "Room Temperature." Add room-temperature pond water to a depth of about 2.5 cm. Label a second culture dish "Cold" and a third "Warm." Add cold water to the second dish and warm water to the third dish. Record the temperature of the water in each of the three dishes.
9. Use a spoon to gently transfer 10 embryos to each dish.
10. Store the dishes according to your teacher's instructions.
11. Use the stereomicroscope to observe your specimens on days 1, 2, 3, 4, and 7. Record any changes in the embryos and make drawings. Remove any injured or dead specimens. Replace the water in each dish with fresh water, always at the same temperature.
12. Clean up your materials and wash your hands before leaving the lab.

Analysis

1. *Summarizing Observations* Summarize your observations.

Summaries will vary according to the data collected.

2. *Evaluating Methods* What temperature is best for culturing frog embryos?

Answers will vary according to the development of the cultures but might state that embryos in the cold culture appear to develop more slowly than those in the room-temperature culture.

Explain why.

The warm culture may develop more quickly but with more fatalities.

3. *Making Inferences* How might the jellylike layer that surrounds the embryo help it survive in water that varies in temperature?

The jellylike layer provides insulation, reducing fluctuations in the embryo's temperature.

The Compound Microscope

Introduction

Have you ever seen the inside of one of your hairs? Can you even imagine doing surgery on the genes inside your cells? The compound microscope is one of the most important instruments used by biologists today. Through observation of microscopic organisms and microscopic structures, scientists have discovered many things about the world around us and how it operates.

In this investigation you will learn the skills necessary to use the microscope correctly. Knowledge of these skills is important to your study of biology.

Materials

- Microscope, compound light
- Lens paper
- Slides, clear and prepared
- Coverslips
- Newspaper
- Scissors
- Medicine dropper
- Beaker, 150 mL
- Forceps

Focus Question

From the information presented in the Introduction and Materials sections above, and in the following **Procedure** section, you should be able to determine the **Focus Question** for this Investigation. For example, a sample Focus Question, given below, has two parts:

1. What are the locations and functions of the compound microscope parts and how are they used to view wet-mounts?

2. What are the differences in high power and low power magnification?

Write these questions on your Vee Form in the section titled "Focus Question."

Knowing Side

With the help of your teacher, your class will discuss the **Subject Area** covered in this investigation. Certain concepts and terminology must be understood before proceeding with the investigation. With the help of your classmates, list these **Concepts** and any **Vocabulary** words unfamiliar to you on your Vee Form. In the **Concept Statements** section of your Vee Form, use these words in sentences that define and explain them.

Procedure

1. Always use both hands when carrying the microscope, placing one hand beneath the *base* with the other hand holding the *arm* of the microscope. Walk with it held close to your body, making sure the electrical cord is wrapped securely around the scope.

Why is this important?

You might run into something and drop the microscope.

Demonstrate the proper carrying technique to your lab partner.

2. Set the microscope at least 5 cm from any table edge.

Why is this important?

It might be knocked off accidentally.

Part A: Use and Care of the Microscope

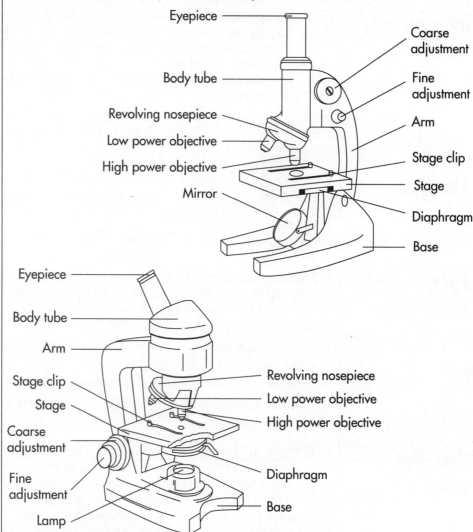

3. If the microscope has a built-in lamp, plug it in and check the position of the cord to be sure it is out of the way. It is possible to get your leg caught in the cord, dragging the microscope with you as you leave the lab table.
4. Begin at the top of the microscope and locate the following parts by comparing your microscope to the appropriate diagram.
5. Locate the *eyepiece* at the top of the *body tube*. The eyepiece contains a lens. If the lens is dirty, clean it with lens paper. Never use anything other than lens paper to clean the lenses of the microscope.

What would be the result of cleaning the lenses with something other than lens paper?

The lens might be scratched.

Locate the number on the eyepiece. The number indicates the magnifying power of the lens in the eyepiece.

What magnification does the lens have? _____10X_____

6. Locate the *revolving nosepiece*, which carries the *objectives*. Locate the shortest objective—this is the *low-power objective.*

What number is written on the low-power objective? _____10X_____

What does this number mean?

The lens magnifies 10 times.

7. Locate the longer objective, or the *high-power objective.*

What number is written on the high-power objective? _____43X_____

What does this number mean?

The lens magnifies 43 times.

Find the total magnification of the microscope at low power and high power by multiplying the number on the eyepiece and the number on the objective. Fill in the chart provided in the **Records** section of the **Doing Side** of your Vee Form.

8. Determine if your microscope has a *lamp* or a *mirror* as a light source. Turn on the light or adjust the mirror using indirect lighting only so that it will reflect light through the hole in the center of the *stage.*

CAUTION: If your microscope has a mirror, never use direct sunlight as a light source. Direct sunlight will damage your eyes.

9. After locating the stage, place a prepared slide selected by your teacher directly over the hole in the stage. Secure the slide with *stage clips.*

Why is this necessary?

to prevent accidental movement of the slide

10. Locate the *coarse adjustment* knob. Turn this knob toward you until the objectives are raised as far as they can be. Using the revolving nosepiece, turn the low-power objective into position over the stage. When focusing the microscope, always begin on low power. While observing the microscope stage from eye level, use the coarse adjustment to move the objective down as far as it will go without touching the slide. Then focus upward with the coarse adjustment until the object to be observed comes into view. **Never focus objectives downward.**

What problems could this cause?

The slide or lens might be cracked or broken.

Complete focusing with the fine adjustment. When the microscope is focused on low power, you may turn the nosepiece to high power. Before switching to high power, make sure that the object you are viewing is exactly in the middle of your field of vision. **Do not use the coarse adjustment to focus at high power.** Focus as before with the fine adjustment until the image is clear.

How does high power change your field of vision?

It decreases it.

How does high power affect the brightness of the image?

The image is less bright.

11. Locate the *diaphragm.* It can be found directly under the stage. Carefully change the setting while looking through the eyepiece.

What is the function of the diaphragm?

It adjusts the amount of light coming through the slide.

What does too much light do to the image? Too little?

It fades the image.

It makes the field too dark to see details.

Part B: Identification of Parts of the Microscope and Their Functions

Complete the table in the **Additional Records and Observation** section on the back of your Vee Form.

Part C: Making a Temporary Wet Mount

1. Clean a glass slide and coverslip with a soft cloth. Be careful to hold both of them by the edges to avoid smudging them with fingerprints.
2. Cut out a lowercase letter *e* from a piece of newspaper.
3. Using a medicine dropper, place one drop of water from the beaker in the middle of the slide.
4. With forceps, place the letter *e* in the drop of water so that it is readable and facing you.

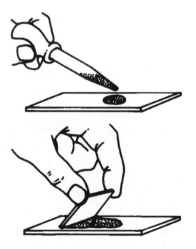

5. Place the coverslip at the edge of the drop of water at a 45-degree angle. Check to see that the water runs along the edge of the cover-slip. Lower the coverslip slowly to avoid air bubbles. There should be a continuous layer of water between the slide and the coverslip. You may need to tap the coverslip gently to eliminate air bubbles. The air bubbles will appear as perfectly round dark areas.

6. As the water evaporates from the slide, you may need to add water to keep your slide fresh. Never lift the coverslip to add or take away water, as this may damage or alter your specimen. Instead, place the tip of the dropper next to the edge of the coverslip and add a drop of water, which will run underneath. This is also the technique to use when adding different strains or solutions to a wet mount. If you add too much water, it may be removed by using the corner of a paper towel to soak up the excess water. If you add stain on one side of the coverslip, you can place a paper towel corner on the other side to draw the stain through.

7. Place your wet mount of the letter *e* on the stage of the microscope. Make sure that the *e* is facing you as you would normally read it.

8. Using low power, focus the microscope on the slide. Review Part A step 10, if needed.

9. Draw and label your observations in the **Additional Records and Observations** section on the back of your Vee Form.

In what three ways is the image of the letter e *different through the microscope as compared to looking at it without the microscope?*

It is upside down, magnified, and backwards.

10. While looking through the eyepiece at the letter *e*, move the slide to your right.

How does the letter appear to move?

It moves to the left.

Now move the slide to the left.

What happens?

It apparently moves to the right.

If you were tracking a microorganism that appeared to be moving from the right side of your field of vision to your left, which way would you move the slide to keep it in view? Why?

to the left; it is really moving toward the right edge

If that same organism suddenly changed direction and started to move toward the bottom of your field of vision, which way would you move the slide to keep it in view? Why?

Move the slide down. The actual movement is in the opposite direction of where

it appears to be.

11. Now observe the letter *e* under high power. Focus following the directions in step 10 of Part A. Draw and label your observation in the Records section of your Vee Form.

12. Dry the microscope completely when you have finished your work. Check to see that the low-power objective is directly over the stage, the objectives are all the way down, and the tube is down as far as it will go. This protects the eyepiece and objectives from possible damage. Wrap the cord around the base and cover the microscope. Store the microscope as directed by your teacher.

13. Clean up your materials and wash your hands before leaving the lab.

Use the information on the **Knowing Side** of the Vee to interpret your results from the **Doing Side**, and then write your **Knowledge Claim**, which should answer your **Focus Question**. One possible **Value Claim** for this investigation might relate to why a biologist chooses different magnifications for viewing a particular specimen. In the Value Claim section of your Vee, write the reasons why a researcher might want to use low power instead of high power in a certain situation, or vice versa.

Going Further

In the appropriate space on your Vee Form write a **New Focus Question** that could be the point of a new investigation based on what you have just learned.

• Make a wet mount of each of the following: fiber from cloth, a colored photograph from the newspaper, and a hair. Draw what you see on each slide using both low power and high power.

• Care and safety need to be emphasized to the students. The more time spent on this investigation, the more success students will have with future investigations using the microscope. Other investigations rely heavily on the premise that students have mastered the basic techniques and have acquired the right attitude toward use and care of the microscope, including safety.

• Caution students to watch carefully as they change to high power, because occasionally slides are thicker than normal. If this is the case, the high-power objective may hit the slide when moved into place.

• Acquaint students with both iris and disc diaphragms and show how to change the setting on each type.

• Warn students not to touch the lamps; they get very hot. If the microscopes have a plastic or vinyl cover, caution students to make sure the lamp is cool before replacing the cover. A hot lamp may melt the cover.

Name _____ Date _____ Class _____

The Compound Microscope

KNOWING SIDE

DOING SIDE

Focus Question:

1. What are the locations and functions of the compound microscope parts and how are they used to view wet mounts?
2. What are the differences in high-power and low-power magnification?

Value Claim:

For viewing a large area of the specimen, low power should be used. However, higher power, while sacrificing a larger area of viewing, allows a magnification of 4X larger. This allows more detailed viewing of the characteristics of a portion of the specimen or a particular cell part.

New Focus Question:

How do other items look under a compound microscope?

Knowledge Claim:

Knowing the locations and functions of the compound microscope enables the viewer to see wet mount specimens at 100X or 430X magnifications.

Concepts:

magnification/power

Vocabulary:

microscope
wet mount
slides
coverslip

Subject Area:

Applied biotechnology

Records:

Magnification Calculation Chart

	Eyepiece	Objective	Total
Low power	10X	10X	100X
High power	10X	43X	430X

Materials:

compound microscope, lens paper, slides (clear and prepared), coverslips, newspaper, scissors, medicine dropper, 150-mL beaker, forceps.

Procedure:

Carry microscope to work area and locate all working parts. Determine their uses. Prepare temporary wet mounts and use learned skills to label drawings of observations.

Concept Statements:

1. Microscopes have many operable parts to enable the viewing of things we cannot see with our naked eye.
2. Temporary wet mounts are made from a specimen laid on a wetted slide covered by a coverslip. They allow the best view when focused under the microscope.

Additional Records and Observations:

Identification of Parts of the Microscope and Their Functions

Part	Function
Eyepiece	magnifies 10X
Body tube	maintains distance between eyepiece and objective
Revolving nosepiece	holds objectives and allows them to move
Low-power objective	magnifies 10X
High-power objective	magnifies 43X (or 40X)
Stage	supports slide
Stage clips	secures slide
Diaphragm	allows the amount of light coming through the stage to be adjusted
Coarse adjustment	moves body tube to allow focusing
Fine adjustment	used to focus and sharpen image
Light source	provides light to view slide
Base	supports microscope
Arm	supports body tube

Crime Lab

Introduction

For best results, Ward's simulated blood should be used for the positive blood slide. (36 M 0018)

Because the microscope enables scientists to see details that ordinarily could not be seen, it is an essential tool for criminal investigations. Many mysteries have been solved by discovering some special detail, one that becomes the evidence supporting a person's guilt or innocence. In this investigation you will be testing your skills as a criminal researcher. The better your technique and observation skills, the better chance you have to solve the crime.

Materials

- Samples of student hair, animal fur (pet), cloth fibers, crystals of baking powder, flour, salt, corn meal, detergent powder, sand
- Medicine dropper
- Coverslips
- Blood slide, positive
- Blood slide, negative
- Microscope, compound light

Focus Question

Design your own **Focus Question** based on the information presented in the Introduction and Materials sections above, as well as the additional information presented in the Procedure section that follows. Write your Focus Question on the Vee Form.

Knowing Side

With the help of your teacher, your class will discuss the **Subject Area** background for this investigation. Then, with the help of your classmates, list the **Concepts** and the new **Vocabulary** words on your Vee Form. In the **Concept Statements** section of your Vee Form, use these words in sentences that define and explain them.

Procedure

In advance, make up crime envelopes with such clues as the "blood" sample and the crystal type so that you will not be rushed to put this all together during lab time.

It is suggested that the teacher make a transparency that will show the way the various clues look under the microscope. This will help students to check their charts for accuracy.

1. Pre-Lab: Before the investigation, you will have an orientation in crime lab work. This training will give you an opportunity to get to know what each type of clue looks like under a microscope. Make wet mounts and look at several strands of hair, fur, and threads from cloth. Look at several crystals of the general types to be used in the lab. The blood slides will be given to you by the teacher.

   Make a drawing on your own paper for each observation that you do. Save this observation sheet; it will be your identification chart for the unknown clues in the upcoming investigation. You should have at least eight drawings in your chart. Do as many drawings as you can.

2. Before you receive your "unknown clues," give your teacher clues such as strands of your hair, threads from the clothes you are wearing, and samples of pet fur or feathers (if you have a pet). Your teacher will place these materials in an envelope along with several crystals and a blood slide. Record the list of your own clues as follows: the color of your hair and whether it is curly or straight, the thread color from your clothes, your pet's fur color and pet type, whether your clothing is made of cotton or a cotton/polyester blend, and any other information your teacher gives you about your envelope. Display this clue list in plain sight next to your work area. (If the person who gets your envelope can identify all the clues, they can link you to the scene of the crime and identify you as the criminal.)

For a positive blood slide, use a prepared slide of human red blood cells. For the negative slide, make a wet mount of a smear of ketchup, red ink, etc.

3. You will receive an envelope containing certain clues that will lead you to your suspect. As the criminal (one of your classmates) was leaving the scene, items such as his or her clothes were caught on a piece of metal, and a few threads, some pet fur, and some crystals were left behind. From your microscopic investigations you might obtain evidence implicating one of your classmates. The forensic team has already isolated the "blood" from the clothes sample. If your sample slide shows cells present, your evidence suggests a possibly violent criminal. If no cells can be seen on low or high power, you have a negative sample and the red fluid is something else that may be important.

4. Observe the clues in the envelope under the microscope at low and high power. Make careful drawings of your observations in the space provided in the **Additional Records and Observations** section on the back of the Vee Form. Summarize the data you have obtained in the **Records** section on your Vee Form. Then scan the clue lists displayed on the desks of your classmates. Narrow them down to the one list of clues that fits your evidence. When you think you have found the suspect for the crime, check with your teacher to see if you are correct.

5. Clean up your materials and wash your hands before leaving the lab.

Use the information on the **Knowing Side** of the Vee to interpret your results from the **Doing Side**, and then write your **Knowledge Claim**. Write a **Value Claim** for this study.

Analysis

Suggested workplaces for suspects, based on clues:
Salt—works near the ocean, perhaps on a boat.
Baking soda—works in a bakery.
Detergent—works in a laundry/cleaning plant
Sand—works in a plant nursery

1. What other things could you examine from the scene of the crime?

bullets, fingerprints

2. What other tools would be useful in a criminal investigation?

computers, chemicals for analyzing substances, resource books

3. Using all your data from your crime sheet:

a. Who would you say committed the crime?

Answers will vary.

b. What kind of establishment did this person frequent?

Answers will vary.

c. Might this have been a crime of violence?

Answers will vary.

Going Further

In the appropriate space on your Vee Form, write a **New Focus Question** that could be the point of a new investigation based on what you have just learned.

Name _____ Date _____ Class _____

Crime Lab

KNOWING SIDE

DOING SIDE

Focus Question:

How can one use microscopy techniques to help convict a suspect in a criminal investigation?

Value Claim:

Microscope work can yield data to solve a crime.

New Focus Question:

How would computers and chemical analyses help convict a suspect?

Knowledge Claim:

By using good microscopy techniques it is possible to isolate data that may be valuable to a criminal investigation.

Records:

Observation summary:
Answers will vary according to envelope contents.

Subject Area:

Forensics

Concepts:

criminal
investigation
observation skills

Vocabulary:

microscopy

Materials:

compound light microscope, blood slides, coverslips, medicine dropper; other materials will vary according to sample envelope contents.

Procedure:

Using a microscope, examine possibly incriminating unknown clues taken from the scene of a crime.

Concept Statements:

1. The microscope can magnify objects unseen with the naked eye.
2. Criminal investigations may be aided by use of microscopy.
3. Better observation skills and mechanical techniques are valuable in researching and solving crimes.

Additional Records and Observations:

Drawings of Hair, Cloth, Pet Fur, Chemical Substance, Blood, etc.

Drawings will vary according to envelope contents.

Designing Control Experiments

Introduction

The scientific method is a series of steps that allows the scientist to establish rational conclusions about natural processes. The steps in the scientific method will help you set up the experiment necessary for this investigation. For an experiment to be a valid one, it must meet two important requirements: (1) only one *variable* is tested and it is the only one that influences the observed results; (2) the experiment is designed in such a way that other scientists can repeat it and get similar results. In this investigation you will have the opportunity to use the scientific method to design control experiments and to carry them out.

Materials

- Petri dishes (2)
- Paper towels
- Water
- Seeds

Other materials will vary according to the hypothesis tested.

Focus Question

HRW material copyrighted under notice appearing earlier in this work.

- Remind students that for an experiment to be valid, only one variable can be manipulated and a control group must be included. Stress the importance of accurate measurement, detailed notes, and observations on a regular basis.
- Discuss the variables in list A and list B. Have students add their own variables to list B. Students will need to formulate at least one hypothesis and decide on the basic design of their experiment before the day of the lab. Stress that they need to plan in advance if they will need materials unavailable at the school.

Scientific research involves asking questions that can be tested and answered by experiment (hypotheses). Often hypotheses are written as "if—then" statements. The more clearly you can write your hypothesis, the easier it will be to design an experiment to test it. Remember, the purpose of the hypothesis is to offer an explanatory answer to your **Focus Question**. Hypotheses are not just educated guesses. There may be many hypotheses or competing alternative explanations you may think of to answer your Focus Question.

1. Choose one variable from each of the two lists below. Write a hypothesis that is testable in the laboratory for each variable. Write this hypothesis as a statement explaining what you think you will find if one variable is changed. Seeds have been chosen for these experiments.

    Example: If seeds are grown with light, then they will not grow as well as seeds grown in darkness.

Main Variables (A)

- Water
- Light/dark
- Nutrients
- Temperature

Other Variables (B)

- Pollutants
- pH (acidity)
- Overcrowding

Hypothesis A

Sample hypothesis: If seeds are planted in nutrient-rich soil, then they will grow better than those in nutrient-poor soil.

_____

Hypothesis B

Sample hypothesis: If seeds are planted in pollutant-free soil, then they will grow better than those in polluted soil.

_____

2. Choose one of these hypotheses. Write a **Focus Question** on the Vee Form that is answered by testing the proposed hypothesis.

Knowing Side

All living things are affected by their environment. Examples of environmental conditions are temperature, light, water, and presence of other organisms. Each condition in the environment is a variable—that is, it may change. You can test the effect of a variable by changing only that variable and keeping all the others the same. When you design an experiment that deals with the change in one variable only, it is called a control experiment.

Why is it important to change only one variable in a control experiment?

You will know if the changed variable is responsible for the observed result.

Why do you think seeds might be good subjects to study in a lab situation?

Seeds are easy to maintain, the environmental conditions are easy to control, and germination is rapid.

With the help of your teacher, your class will discuss the **Subject Area** covered in this investigation. Then, with the help of your classmates, list the **Concepts** and the new **Vocabulary** words dealing with both the scientific method and seeds on your Vee Form. Now use these words in sentences that define and explain them in the **Concepts Statements** section of your Vee Form.

Procedure

3. In the **Procedure** section of the Vee, describe or draw a sketch showing how you would set up your experiment to test your chosen hypothesis. Remember to use only the materials available.

Which group is considered the experimental group? Why?

Answers will depend on individual experiments.

Which group is considered the control group? Why?

Answers will depend on individual experiments.

4. Set up your experiment according to the design in your procedure. Use the materials specified and follow as many of the procedure steps as you can on the first day.

5. Set up a log in which you can record your observations. These should be dated, done at least every other day, and should include both drawings and written explanations of your observations. Record these observations in the **Additional Records and Observations** section on the back of your Vee Form. The more records you have, the better your knowledge claims will be. You may also wish to present these data in charts, tables, or graphs.

6. Clean up your materials and wash your hands before leaving the lab.

Use the information on the **Knowing Side** of the Vee to interpret your results from the **Doing Side**, and then write your **Knowledge Claim**. Remember that the Knowledge Claim evaluates the hypothesis as well as the results of your experiment. Be sure to include any errors made during the investigation that might have affected the results. Write a **Value Claim** for this study.

Analysis

1. How do your records support your original hypothesis? If they do not, what do you think happened?

Answers will depend on individual experiments.

2. Even if your records seem to support your hypothesis, how can you increase the likelihood that they do?

Retest several times; make sure that similar experimental and control groups produce the same results.

3. What was the most difficult part of this investigation?

Answers will depend on individual experiments.

4. What other variables might be important when considering conditions in a seed's natural environment?

Sample Answer: predators, competition from different plants, and duration and intensity of light

5. How did you ensure that only the variable in the hypothesis was being tested?

by use of a control group

Going Further

In the appropriate space on your Vee Form, write a **New Focus Question** that could be the point of a new investigation based on what you have just learned.

• Devise a hypothesis and experiment to test your New Focus Question. See if you can perform the experiment under the best possible controlled conditions. Be careful to keep all the variables the same, except the one being tested.

Designing Control Experiments

KNOWING SIDE

DOING SIDE

Focus Question:

Answers will vary. Sample answer: How will lighting conditions affect the sprouting of seeds when all other variables are kept constant?

New Focus Question:

Sample: Will seeds sprout as well in the dark if placed in a cold environment?

Subject Area:

Scientific Method
Ecology
Plant Physiology

Concepts:

variable

control experiment

environmental conditions

Vocabulary:

hypothesis

Concept Statements:

1. A control experiment keeps all factors constant except the one variable under consideration.
2. Hypotheses are testable statements that may be confirmed or refuted by experimental results.
3. Control experiments are designed to be repeatable by others.
4. Environmental conditions affect the growth and functions of living things.

Value Claim:

Answers will vary. Sample answer: Some seeds that sprout best in the dark must be somewhat buried to best germinate. An example would be grass seed, which must be raked into the soil after scattering for it to sprout most effectively.

Knowledge Claim:

Answers will vary. Sample answer: Seeds kept in the dark will sprout more quickly than those kept in the light.

Records:

[See Additional Records and Observations on the following page]

Materials:

petri dishes, paper towels, water, seeds; other materials will vary.

Procedure:

Answers will vary. Sample answer: Place seeds of the same size and type in closed petri dishes with moistened paper. Place half in the dark and half in the light and every day for five days record how many in each of the two groups sprout.

Diversity of Life

Introduction

A class discussion of the organization of living things from atoms to biomes would be a good extension of this investigation.

Materials

If chemically preserved specimens are used, they should be sealed in clear plastic jars. Students should not open them.

Focus Question

Knowing Side

Procedure

Any easily preserved specimen will suffice. Especially good choices would be a tapeworm and a mollusk. The microorganisms used in the investigation can be algae, yeast, bacteria, or protozoa. Encourage students to bring their own live specimens, or to observe live specimens outside.

What are some differences between you and your classmates? between you and the plants and animals in your classroom? between you and a rock? In this investigation you will observe many characteristics that set living things apart from nonliving things. You will have the opportunity to gather data through careful observation of many different kinds of living things. You will then compare the similarities and differences of these organisms, and their relationships to one another and to their environments.

- Preserved specimens (2)
- Live specimens (2)
- Live specimen, microscopic (2)
- Metric ruler
- Microscope, compound

- Slides (3)
- Medicine dropper
- Coverslips (3)
- Paper towels
- Containers for live specimens (2)

Design your own **Focus Question** based on the information presented in the Introduction and Materials sections above, as well as the additional information presented in the Procedure section that follows. Write your Focus Question on the Vee Form.

With the help of your teacher, your class will discuss the **Subject Area** covered in this investigation. Then, with the help of your classmates, list any **Concepts** and new **Vocabulary** words that relate to the characteristics of living things on your Vee Form. Now use these words in sentences that define and explain them in the **Concept Statements** section of your Vee Form. These Concept Statements should list what you already know about the characteristics of living things.

1. Choose two preserved specimens, two living macroscopic specimens, and two living microscopic specimens to study. In the chart provided in the **Additional Records and Observations** section on the back of the Vee Form, draw a small sketch of each organism. Describe the characteristics of living things listed in the chart for each of the specimens as completely as possible.
2. Measure and plot the size of each of your specimens on the bar graph provided. It will allow you to compare specimens and to see the great diversity in the sizes of living things.
3. Add three more specimens to the size-range graph. Choose specimens you could not bring into the classroom, for example, a whale. You may use reference books to determine the sizes of these specimens.
4. Clean up your materials and wash your hands before leaving the lab.

Use the information on the **Knowing Side** of the Vee to interpret your results from the **Doing Side**, and then write your **Knowledge Claim**. Write a **Value Claim** for this study.

Analysis

1. Which question(s) in the table was the most difficult to answer from observation alone? Why?

Reproduction and adjustment to the environment, for example. These were difficult to see during the allotted time.

2. What other characteristics of living things were not explored in this lab?

Sample Answers: adaptation to new environments, cellular organization, natural habitat, living habits

Going Further

In the appropriate space on your Vee Form write a **New Focus Question,** that could help you gain a better understanding of the similarities and differences in living things, building on what you have just learned.

• Choose one specimen that you have observed and describe how its special structure has enabled it to live successfully in its natural environment. Provide your specimen with an environment that closely models its natural habitat. Make your own Vee Form for this study, including records and drawings from your observations. Write your Knowledge and Value Claims on your form.

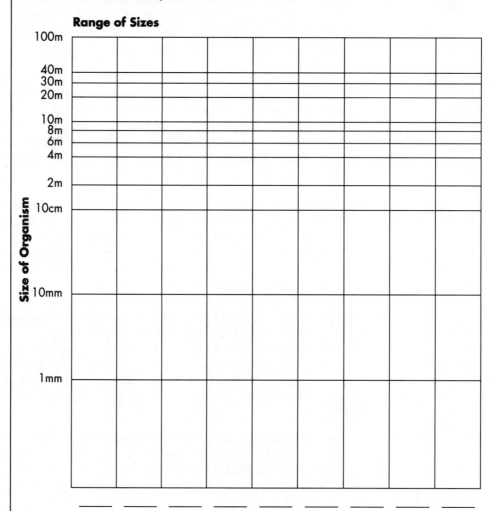

Range of Sizes

Size of Organism

100m
40m
30m
20m
10m
8m
6m
4m
2m
10cm
10mm
1mm

Type of Organism

Name _____ Date _____ Class _____

Diversity of Life

KNOWING SIDE

Subject Area:

Biological variation
Evolution
Cell theory

Concepts:

cells
life span
environment
adaptation
reproduction

Vocabulary:

metabolism
stimuli
unicellular
multicellular

Focus Question:

What characteristics that set living things apart from nonliving things can be seen in our specimens? What are the similarities and differences among our specimens?

New Focus Question:

If I could provide one of my specimens with a suitable living environment, what more might I learn about them?

Concept Statements:

All organisms:
1. are made up of cells—unicellular or multicellular.
2. are organized—raw materials are highly organized into more complex substances.
3. use energy—metabolism is the sum of all the ways in which an organism uses energy.
4. have a life span—beginning, growth, maturity, decline, and death.
5. reproduce—mature organisms create new organisms by either sexual or asexual means.
6. respond to stimuli—react to conditions in the environment.
7. adapt to their environment—change in ways that often make them better suited to their environment.

DOING SIDE

Value Claim:

Due to variations in habitats, organisms show great diversity in cell organization, living habits, life span, size, etc. Knowing more about organisms and their habitats will enable humans to better protect their environments.

Knowledge Claim:

Living things use energy to move and to respond to stimuli.

Records:

[See Additional Records and Observations on the following page]

Materials:

preserved specimens (2), live macroscopic specimens (2), live microscopic specimens (2), compound microscope, slides (2), medicine dropper, coverslips (2), paper towels, containers for live specimens (2).

Procedure:

Study six specimens, drawing and describing each according to the characteristics of living things they show. Compare the sizes of specimens and illustrate them by plotting them on a bar graph.

Additional Records and Observations:

Organism		Unicellular or Multicellular	Type of Metabolism	Evidence of Life Span (growth)	Form(s) of Reproduction	Response to Stimuli	Adjustment(s) to Environment	Adaptations
Sketch								

Cytoplasm and Organelles

Introduction

Many cell organelles have been discovered through the use of electron microscopes. However, certain organelles of the cell can be seen with the compound microscope. This investigation will enable you to observe the activities of the cytoplasm and to observe some of the organelles found in the cytoplasm of living *Elodea* and *Amoeba* cells.

Materials

- *Amoeba* culture
- *Elodea*
- Slides (2)
- Coverslips (2)
- Microscope, compound light
- Medicine dropper

Focus Question

Design your own **Focus Question** based on the information presented in the Introduction and Materials sections above, as well as the additional information presented in the Procedure section that follows. Write your Focus Question on the Vee Form.

Knowing Side

With the help of your teacher, your class will discuss the **Subject Area** covered in this investigation. Then, with the help of your classmates, list any **Concepts** and new **Vocabulary** words that relate to the characteristics of living things on your Vee Form. In the **Concept Statements** section of your Vee Form, use these words in sentences that define and explain them.

Procedure

1. Remove a drop of *Amoeba* culture from the bottom of the culture container and place it on a slide. Add a coverslip. Find an *Amoeba* under low power. Observe the cytoplasm.

Is there movement in the cytoplasm? If so, how does this movement take place?

Yes. A temporary cytoplasmic projection is pushed out. Some of the cytoplasm flows into this

projection.

You may wish to order the following from Ward's.

Amoeba Culture Kit
(87 M 0391)

Elodea may also be obtained from them, or from a pet store.
(86 M 7500) Bunch of 10

Locate a vacuole, a bubblelike structure in the cell. Switch to high power and observe the vacuole. This structure may store water, food, or waste material.

Did each Amoeba *have at least one vacuole?*

Yes, although occasionally students will have difficulty identifying one.

Although the *Amoeba* is not an animal, it does have the characteristics of a typical animal cell. You may wish to point out the differences between plant and animal cells at this time.

2. In the **Additional Records and Observations** section on the back of your Vee Form, draw the *Amoeba* under low and high power in the spaces provided and label the organelles that you have observed.
3. Obtain a leaf from the top whorl of leaves on a sprig of *Elodea*. Make a wet mount of the leaf. Observe it under low power. Locate the chloroplasts within a cell.

Are the chloroplasts uniformly distributed or near the edge of the cell?

They are located at the edge of the cell.

What function do these organelles have?

photosynthesis

Switch to high power. Observe one cell carefully for several minutes. Describe any motion of the chloroplasts.

Movement is in a circular path; all movement is in the same clockwise or counterclockwise direction.

Locate a vacuole. *How many are there?* _____ one large vacuole

4. Draw the *Elodea* cell under low and high power in the spaces provided in the **Additional Records and Observations** section on the back of your Vee. Label the organelles that you observed.
5. Compare the *Amoeba* and *Elodea* cells by completing the table in the **Records** section of your Vee Form.
6. Clean up your materials and wash your hands before leaving the lab.

Use the information on the **Knowing Side** of the Vee to interpret your results from the **Doing Side**, and then write your **Knowledge Claim**. Also write a **Value Claim** for this study.

Analysis

1. Does the cytoplasmic movement serve the same function in both organisms?

   No. In the *Amoeba* the cytoplasmic movement moved the organism through the environment; in the *Elodea* it moved the chloroplasts and other materials throughout the cell.

2. Compare the size and number of vacuoles found in the *Amoeba* and *Elodea* cell.

   In *Elodea* there is one large vacuole. In the *Amoeba* there are usually several small vacuoles.

3. Formulate a hypothesis to explain the difference in vacuoles in the *Elodea* cell and the *Amoeba*.

   Answers will vary.

Going Further

In the appropriate space on your Vee Form write a **New Focus Question** on your Vee Form that could be the point of a new investigation based on what you have learned.
• Observe other organisms, such as *Euglena* and *Paramecium*. Make a comparison of the organelles used for locomotion.

Name _____ **Date** _____ **Class** _____

Cytoplasm and Organelles

KNOWING SIDE

DOING SIDE

Focus Question:

How do cytoplasm and organelles differ between *Amoeba* and *Elodea* when seen under the microscope?

Value Claim:

All cells are not identical. Variation occurs across phyla.

New Focus Question:

1. Are any other protists similar to plants? (*Euglena? Paramecia?*)
2. Do one-celled organisms have different organelles of locomotion?

Knowledge Claim:

The cytoplasm appears more grainy in the *Amoeba*, while there are many chloroplasts and a large vacuole in *Elodea*. Thus it is harder to see its cytoplasm and make a comparison. The *Elodea* plant has chloroplasts, a cell wall, and a large water vacuole, all missing in the *Amoeba*.

Records:

	Amoeba	Elodea
Chloroplasts		✓
Nuclei	✓	✓
Cell membrane	✓	✓
Cell wall		✓
Vacuoles	✓	✓
Cytoplasm	grainy	too full to observe

Subject Area:

Cell theory

Materials:

Amoeba culture, *Elodea*, slides, coverslips, microscope, medicine dropper.

Vocabulary:

organelle
cytoplasm

Procedure:

Observe *Amoeba* and *Elodea* under the microscope and observe their cytoplasm and organelles.

Concepts:

cell
protist
plant

Concept Statements:

1. Microscopes are used to view cell cytoplasm and organelles.
2. *Amoeba* is a protist.
3. *Elodea* is a plant.
4. Cells are the basic unit of structure and function in all living things.
5. Organelles within the cytoplasm have different jobs in different cells.

Additional Records and Observations:

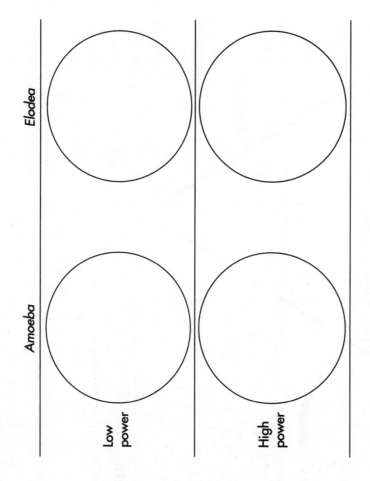

Amoeba

Elodea

Low power

High power

Structure and Function of Cells

Introduction

You may wish to show diagrams of frog anatomy found in the text (see Tour of a Frog on page 563), or other reference material.

Materials

This will enable students to understand more about the parts of the frog's body from which the cells were obtained.

Focus Question

Knowing Side

Procedure

The following supplements to this lab are available from Ward's.

Cell Structure and Function (170 M 9109) color slides

The Cell (75 M 0110) transparencies

Analysis

One of the basic concepts of biology is that structure is related to function at many different levels of organization. In this investigation you will investigate this connection at the level of the cell. Cells are the basic unit of structure and function of all living things. In this investigation, it will be your job to relate function to structure in three types of typical animal cells.

- Microscope, compound light
- Lens paper

- Prepared slides:
  frog sperm
  frog blood
  frog skin

Design your own **Focus Question** based on the information presented in the Introduction and Materials sections above, as well as the additional information presented in the Procedure section that follows. Write your Focus Question on the Vee Form.

With the help of your teacher, your class will discuss the **Subject Area** background for this investigation. Then, with the help of your classmates, list the **Concepts** and the new **Vocabulary** words on your Vee Form. In the **Concept Statements** section of your Vee Form, use these words in sentences that define and explain them.

Examine each of the three frog cell slides separately. In Figure 1 of the **Additional Records and Observations** section on the back of your Vee Form, sketch each cell as viewed at both low and high power.

Which type of cell has a nucleus?

all have one

Describe the differences in shape and cellular content, in these three cell types in Table 1 provided in the **Additional Records and Observations** section on the back of your Vee Form.

Use the information on the **Knowing Side** of the Vee to interpret your results from the **Doing Side**, and then write your **Knowledge Claim**. Write a **Value Claim** for this study.

1. The main function of a cell is to maintain its own life processes. How does a nucleus allow a cell to accomplish this function?

The nucleus controls the life activities of the cell.

The prepared slides used in this Lab are available from Ward's.
Frog Sperm
(92 M 8803)
Frog Blood
(92 M 3640)
Frog Skin
(92 M 3643)

2. Compare the shapes of the three types of cells observed.

Skin cells are flat and broad. Blood cells are small, round, and smooth. Sperm cells have a head and a tail.

3. Which cell was best adapted for lining or covering? What characteristics let it serve this function well?

Skin cells are shaped to overlap and cover; they are flat and broad.

4. Which cell was best adapted for carrying oxygen through the blood vessels? What characteristics let it serve this function well?

Blood cells are shaped to pass through vessels, being small, round, and smooth.

5. Which cell adapted for traveling to and joining with the egg cell? What characteristics let it serve this function well?

The tail of a sperm cell is designed for locomotion, and the head for penetration.

6. With a sketch or in words, design a cell that would be adapted to take in nutrients from foods.

Answers will vary but should indicate a cell that has a large surface area and many nutrient-holding vacuoles.

Going Further

Look at muscle or nerve cells in slides or photographs and relate their functions to their structures.

Name _____ Date _____ Class _____

Structure and Function of Cells

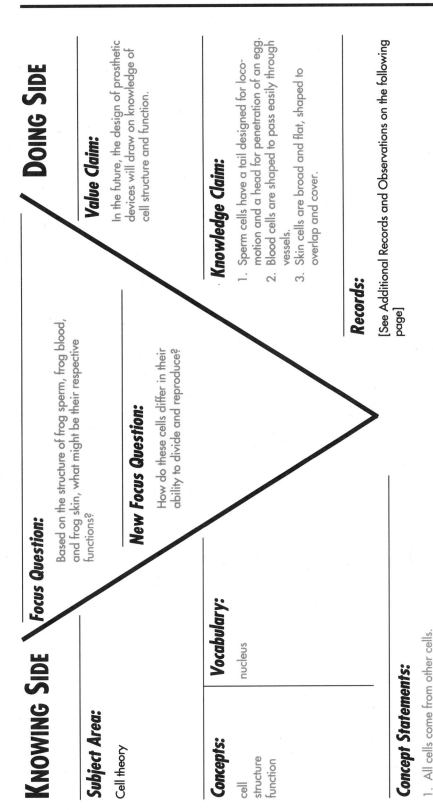

KNOWING SIDE

DOING SIDE

Focus Question:

Based on the structure of frog sperm, frog blood, and frog skin, what might be their respective functions?

New Focus Question:

How do these cells differ in their ability to divide and reproduce?

Subject Area:

Cell theory

Concepts:

cell
structure
function

Vocabulary:

nucleus

Concept Statements:

1. All cells come from other cells.
2. Cells are the basic unit of all living things.
3. Cell nuclei direct the cell functions.

Value Claim:

In the future, the design of prosthetic devices will draw on knowledge of cell structure and function.

Knowledge Claim:

1. Sperm cells have a tail designed for loco-motion and a head for penetration of an egg.
2. Blood cells are shaped to pass easily through vessels.
3. Skin cells are broad and flat, shaped to overlap and cover.

Records:

[See Additional Records and Observations on the following page]

Materials:

compound light microscope, lens paper, prepared slides (frog sperm, frog blood, frog skin)

Procedure:

Observe three different cell types of the frog and sketch them at low and high powers. Label nuclei and describe the differences among these three cells in content, shape, and independent movement.

Additional Records and Observations:

Table 1 Description of Cell Differences

	Frog Sperm	Frog Blood	Frog Skin
Shape			
Content			

Figure 1 Frog Cell Sketches

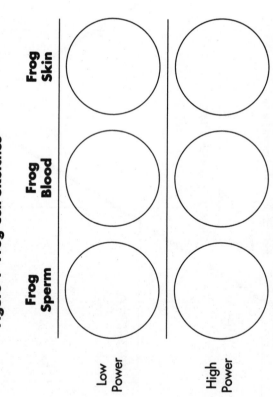

Diffusion and Cell Membranes

Introduction

Not all chemicals can pass through a cell membrane with equal ease. Chemicals move from areas of high concentration to areas of low concentration as they pass in and out of a cell. What regulates this passage of substances diffusing in and out of a cell? Sometimes scientists use models to help answer such a difficult question. This investigation will use a model of a living cell to show changes that are controlled by the cell membrane.

The cell membrane determines which substances can diffuse into or out of a cell. Each cell membrane thus has its characteristic *permeability.* Some chemicals can pass through the cell membrane, but others cannot.

In this investigation you will use an egg with the shell dissolved as a model for a living cell membrane. You will then predict the results of an experiment involving the movement of materials through a membrane.

Materials

- Fresh eggs (2)
- corn syrup
- Vinegar
- Balance
- 250 mL beakers (2)
- 400 mL or 600 mL beakers (2)
- Marking labels & a marking pen

Focus Question

Design your own **Focus Question** based on the information presented in the Introduction and Materials sections above, as well as the additional information presented in the Procedure section that follows. Write your Focus Question on the Vee Form.

Knowing Side

With the help of your teacher, your class will discuss the **Subject Area** background for this investigation. Then, with the help of your classmates, list the **Concepts** and the new **Vocabulary** words on your Vee Form. Now use these words in sentences that define and explain them in the **Concept Statements** section of your Vee Form.

Procedure

The investigation will take 2–3 days due to two 24-hour waiting periods. This can be done as a demonstration with students involved in the set up. It is less costly as a demonstration.

1. Mark one large beaker "Water" and the other beaker "Syrup." Determine the mass of each egg to the nearest 0.1 g and record them in the **Additional Records and Observations** table on the back of your Vee Form.
2. Pour 200 mL of vinegar solution into each large beaker. Place each egg into the appropriate beaker and place a 250-mL beaker with 100 mL of water over each egg to keep it submerged. (Add more vinegar solution if the egg is not covered by the vinegar.) Pour some out if the vinegar runs over when the 250-mL beaker is placed over the egg. Store your beakers for 24 hours in the area specified for your class.
3. After 24 hours, observe the eggs and record your observations in the Vee table. Pour the vinegar in the sink, remove the eggs, rinse them with water, and place them on paper towels labeled "Water" for the water egg and "Syrup" for the syrup egg. Determine the mass of each egg and record them in the table on the back of your Vee Form.

4. Place the "Syrup" egg in the "Syrup" beaker and add syrup until the egg is covered. Place the "Water" egg in the "Water" beaker and add distilled water until the egg is covered. Store the beakers with the eggs in the same location for another 24 hours.

5. Predict how the mass of each egg is going to change in the next 24 hours. Keep in mind that the egg is covered with a membrane, that the inside of the egg is made of egg white (water and dissolved protein), and that egg yolk is mainly fat and water. Syrup is sugar dissolved in water. Record your predictions in the table on the back of your Vee Form.

6. Observe the eggs after the allotted time has passed. Record your observations in the table. Include the final masses of the eggs.

Use the information on the **Knowing Side** of the Vee to interpret your results from the **Doing Side**, and then write your **Knowledge Claim**. Write a **Value Claim** for this study.

Analysis

1. What effect did the vinegar have on the egg?

   The vinegar dissolved the eggshell.

2. How did the results of this investigation compare with your prediction?

   Answers will vary depending on the students' predictions.

3. What kind(s) of material(s) apparently moved through the egg membrane?

   Water moved through the egg membrane from the vinegar solution and from the interior of the egg itself to the outside.

4. What is the movement of water across a membrane?

   Diffusion of water through a membrane is called osmosis.

5. Which egg was in a hypertonic solution? Explain.

   The water egg was in a hypertonic solution because water diffused into the egg, increasing the egg mass.

6. Which egg was in a hypotonic solution? Explain.

   The syrup egg was in a hypotonic solution because vinegar solution diffused from the egg into the syrup solution, decreasing the egg mass.

Going Further

In the appropriate space on your Vee Form, write a **New Focus Question** that could be the point of a new investigation based on what you have just learned.

Name _____ Date _____ Class _____

Diffusion and Cell Membranes

KNOWING SIDE

DOING SIDE

Focus Question:

How do models and living cells illustrate the properties of diffusion, osmosis, and permeability through a cell membrane?
What materials will move across the membrane of an egg cell in hypertonic and hypotonic solutions?

New Focus Question:

Could the same results be obtained with living yeast cells and a color dye solution that would not kill them?

Value Claim:

It is important that cells keep out substances that could harm them; it is also important for them to eliminate waste substances. This is accomplished by diffusion through the cell membrane.

Knowledge Claim:

Water passes freely across the membrane and is a hypertonic solution to the egg. Syrup is a hypotonic solution to the egg, and vinegar water diffuses out of the membrane until equilibrium is reached.

Records:

[See Additional Records and Observations on the following page]

Subject Area:

Cell theory
Diffusion

Vocabulary:

Concepts:

cell
membrane
model
equilibrium

osmosis
diffusion
permeability
isotonic
hypotonic
hypertonic

Materials:

2 eggs, corn syrup, vinegar, balance, 250-mL beakers (2), 400-mL or 600-mL beakers (2), marking labels.

Procedure:

Determine the mass of each egg and soak them in vinegar overnight. Rinse them, and put one egg in water and one in syrup. Observe after 24 hours.

Concept Statements:

1. Cells are the basic unit of structure and function of all living things.
2. Substances travel from an area of greater to lesser concentration in diffusion until equilibrium is reached.
3. Osmosis is the diffusion of water through a semipermeable membrane.
4. Cell membranes are permeable to certain substances and not to others.
5. A cell placed in an isotonic solution has the same solute concentration inside and outside the cell.
6. A cell placed in a hypotonic solution has a higher solute concentration than that of the environment.
7. A cell placed in a hypertonic solution has a lower solute concentration than that of the environment.

Additional Records and Observations:

Prediction (will vary)

Mass of Fresh Egg with Shell	Observation after 24 Hours in Vinegar	Mass after 24 Hours in Vinegar	Observation after 24 Hours in H₂O or Syrup	Final Mass of Egg
"Water Egg" X	Shell is dissolved	X + increase	Egg is very full Increase in mass	greater
"Syrup Egg" X	Shell is dissolved	X + increase	Egg looks smaller membrane is not tight	less

Plant and Animal Interrelationships

Introduction

Oxygen and carbon dioxide are gases used and released by living organisms. Animals and plants use oxygen for cellular respiration and give off carbon dioxide, but plants also use carbon dioxide for photosynthesis. In this investigation you will be setting up closed mini-systems containing plants and animals to find out if there are any observable interrelationships between them. Snails and *Anacharis* live well in a pond or an aquarium, so you will work with them.

Indicators are dyes used to show the presence of certain chemicals by their color change. Bromothymol blue (BTB) is an indicator that turns green or yellow in the presence of a weak acid. Carbon dioxide forms a weak acid in water, so bromothymol blue can indirectly indicate the presence of CO_2.

Materials

A nonalcoholic BTB solution is available from Ward's. (38 M 9100)

- Small containers with lids or stoppers (15 mm x 60 mm, for example) (8)
- Labels or glass-marking pencil
- Small water snails (4)
- Pieces of *Anacharis* or other aquarium plant (4)
- Bromothymol blue solution (nonalcoholic)
- Dechlorinated tap water
- Light source
- White paper

Focus Question

Design your own **Focus Question** based on the information presented in the Introduction and Materials sections above. If you need additional information, you may want to read the Procedure section as well. Write your Focus Question on the Vee Form.

Knowing Side

With the help of your teacher, determine the **Subject Area** that applies to this lab, and write this on your Vee. Then, with the help of your classmates, list any **Concepts** and new **Vocabulary** words that apply to this investigation. List in your **Concept Statements** section any general statements you can make about plants, animals, and bromthymol blue, and how they might be related.

Procedure

If small uniform containers are not available, prescription bottles, baby food jars, or other items can be substituted. Screw cap test tubes also work very well.

Snails and *Anacharis* may be purchased in any aquarium supplies store.

Dechlorinated tap water may be obtained by allowing the water to sit in a container with a large surface area overnight, allowing the chlorine to evaporate. Commercial dechlorinating agents may be purchased from pet stores as well.

CAUTION: Bromothymol blue is a skin irritant. Put on an apron, goggles, and gloves during the investigation.

1. Label 8 containers 1A, 1B, 1C, and 1D for set #1 and 2A, 2B, 2C, and 2D for set #2. Add dechlorinated water to each container, leaving some space at the top of each container.
2. To each container add 4 drops of BTB solution. You may now seal 1A and 2A, as nothing more will be added to them.
3. Add one snail each to containers 1B and 2B.
4. Add a leafy stem portion of *Anacharis* to containers 1C and 2C.
5. Add one snail and a leafy stem portion of *Anacharis* to 1D and 2D. Be sure all caps/lids are watertight.

Which container in each set is the control? _____ "A" of both sets _____

6. When all containers are watertight, place set #1 (A through D) under strong artificial light and place set #2 (A through D) in a dark closet or box.

7. Observe both sets after 24 hours. Place a piece of white paper behind the containers to facilitate color determination. Record BTB color and the condition of the organisms in the **Additional Records and Observations** section on the back of your Vee Form.
8. Switch the environmental conditions so that set #1 is now in the dark and set #2 is in the light.
9. Wait 24 more hours and record your observations again on the back of your Vee.
10. Clean up your materials and wash your hands before leaving the lab.

Use the information on the **Knowing Side** to interpret your results from the **Doing Side**, and then write your **Knowledge Claim**. Write a **Value Claim** for this study.

Analysis

1. Since both organisms live well in a fish tank, why might they suffer in this experiment?

   Accumulation of CO_2 and O_2 is one explanation.

2. Why were containers 1A and 2A used even though no organisms were placed in them?

   In order to compare the color changes in the other containers to the initial colors.

3. If the indicator had changed color in containers 1A and 2A, how could you explain this?

   Small organisms in the water might have given off enough CO_2 to make the control containers acidic.

Going Further

You may wish to order the following items from Ward's.

Aquatic Snails
87 M 4350 (Set of 25)

Anacharis
86 M 7500 (Bunch of 10)

Dechlorinating Agent
21 M 2292 (Superchlor)

Finish your Vee Form by writing a **New Focus Question** for a possible experiment suggested by the results of this study.

The narrow pH range of sensitivity for BTB is 6.0 (yellow) to pH 7.6 (blue). Give your stock solution enough BTB to give each container a light blue tint to begin. A 0.1% stock solution is made by dissolving 0.5 g BTB powder in 500 mL distilled water. Add to this, drop by drop, a dilute solution of ammonium hydroxide until the solution turns blue. If your tap water is alkaline, the ammonium hydroxide will not be necessary. Do not use a purchased BTB solution from supply houses that is not aqueous. It normally contains alcohol and will kill the organisms.

If you notice some strange things happening, seize the moment to discuss them. For example, the D containers might show a green/yellow color at the top and a blue tint at the bottom. The snail might be found at the top and the plant may be resting at the bottom.
Green plants and animals both undergo cellular respiration continuously. While in the light, plants almost immediately use that released CO_2 in photosynthesis, so it does not build up in the container. In the dark with no photosynthesis, BTB turns green with the accumulation of CO_2. As the CO_2 accumulates, it forms a weak carbonic acid with the water, and BTB responds with a change in color. When moved into the light, as photosynthesis begins again, CO_2 is used more rapidly than it is released as the plant pulls the CO_2 from the water and the indicator turns back to blue.

Name _____ Date _____ Class _____

Plant and Animal Interrelationships

KNOWING SIDE

DOING SIDE

Subject Area:

Cell theory
Homeostasis
Gas exchange

Focus Question:

What does bromothymol blue indicate in various combinations of plants and animals in sealed containers placed in the dark and in the light?

Value Claim:

We had better do all we can to protect habitat for plants, since we cannot live without their O_2 release. Many plants could live without animals.

New Focus Question:

What is the right balance of plants and animals for an aquarium?

Knowledge Claim:

BTB shows that while plants are photosynthesizing, they use animal produced CO_2. When there is no light and reduced photosynthesis, the CO_2 level rises.

Records:

[See Additional Records and Observations on the following page]

Concepts:

photosynthesis
cellular respiration

Vocabulary:

O_2 (oxygen)
CO_2 (carbon dioxide)
Bromothymol blue

Materials:

small containers (8), labels, small water snails (4), Anacharis, bromothymol blue, dechlorinated tap water, light source, white paper.

Procedure:

Prepare 8 sealed containers: 1A & 2A: BTB and H_2O; 1B & 2B: BTB, H_2O, and snail; 1C & 2C: BTB, H_2O, and Anacharis; 1D & 2D: BTB, H_2O, snail, and Anacharis. Place set #1 in light, #2 in dark, and observe after 24 hours. Switch light conditions for both sets and observe after 24 more hours.

Concept Statements:

1. Plants use O_2 for cellular respiration and CO_2 for photosynthesis.
2. Animals use O_2 for cellular respiration and give off CO_2 as a waste product.
3. Indicators are dyes that change color in the presence of certain chemicals.
4. Bromothymol blue turns green or yellow in the presence of a weak acid (HCO_3) made by CO_2 and H_2O.
5. Plants need CO_2 and sunlight to photosynthesize and produce O_2 as a byproduct.
6. Cellular respiration breaks down glucose. Photosynthesis produces glucose.

Additional Records and Observations:

After 24 Hours:

	Light	Dark	
1A	No change	No change	**2A**
1B	Green	Green	**2B**
1C	Blue	Blue	**2C**
1D	Blue/organisms thriving	Green/organisms thriving	**2D**

After Switched Conditions and 48 Hours:

	Dark	Light	
1A	No change	No change	**2A**
1B	Green	Green	**2B**
1C	Lt. blue to green	Blue	**2C**
1D	Blue/snail may be dead	Blue/organisms thriving	**2D**

Release of Energy

Introduction

One chemical made by plants to store the sun's energy is the carbohydrate sucrose. It is a disaccharide made from two monosaccharides. All living things must have energy to power their cells. If, for some reason, these cells stop functioning, even for a few minutes, they will cease to live. If enough of them stop carrying on the life processes, the whole organism will die. In this investigation you will have a chance to observe energy release by an organism and see the evidence of the tremendous power contained in the bonds of the sucrose molecule. In Part A, you will examine energy production in a partially aerobic sucrose-yeast solution. In Part B, you will observe normal anaerobic fermentation energy production in a molasses-yeast solution.

Based on your data, you will try to determine whether both cellular respiration and fermentation have occurred in the sucrose-yeast solution.

Materials

The molasses solution may be made by adding 50 mL molasses to 1L of water.

- Vacuum bottles, 500 mL (2)
- Two-hole stoppers for vacuum bottles (each with one thermometer and one 10-cm piece of glass tubing inserted)
- Molasses solution, 750 mL
- Limewater, 150 mL
- Flask, 250 mL (1)
- Dried yeast package
- Sucrose, 75 grams

Focus Question

Design your own **Focus Question** based on the information presented in the Introduction and Materials sections above, as well as the additional information presented in the Procedure section that follows. Write your Focus Question on the Vee Form.

Knowing Side

With the help of your teacher, your class will discuss the **Subject Area** background for this investigation. Then, with the help of your classmates, list the **Concepts** and the new **Vocabulary** words on your Vee Form. In the **Concept Statements** section of your Vee Form, use these words in sentences that define and explain them.

Procedure

Part A: Sucrose-Yeast Energy Production

1. Set up your one vacuum bottle according to the diagram below.

2. Mix 75 grams of sucrose in 400 mL of water.
3. When the sucrose has dissolved, add one-half package of fresh yeast and stir.
4. Pour the sucrose-yeast solution into a vacuum bottle until it is approximately three-fourths full.
5. Adjust the thermometer so that it extends down into the sugar-yeast solution.

Record the temperature of the solution in the table in the **Additional Records and Observations** section on the back of your Vee Form. Continue to record the temperature as often as possible during the next two days. You and your lab partner may take turns during the school day and take temperature data between class periods.

Prepare a graph of your observations on the back of your Vee Form. Plot temperature by putting a point on the graph that corresponds to the temperature and the time. Complete the graph by drawing a curve through the plotted points. The curve will show the yeast's energy production.

Part B: Molasses-Yeast Energy Production

If Part B will be done at the same time as Part A, a third vacuum bottle setup will be necessary.

1. Label the other vacuum bottles #1 and #2. Pour molasses solution into both vacuum bottles until they are three-fourths full.
2. Add one-half of a package of dried yeast to vacuum bottle #1, but NOT to #2.

What purpose will the vacuum bottle without the yeast serve?

It is the control for the experiment.

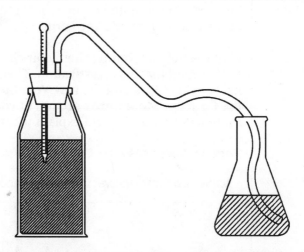

At this point, have the students use the lime-water to test for the presence of CO_2. It turns cloudy in the presence of carbon dioxide.

3. Stopper the vacuum bottles, as shown above, in such a way that the thermometer bulb is in the molasses solution and the end of the glass tubing is above the molasses solution. Attach a piece of rubber tubing to the glass tubing protruding from each stopper. Insert the opposite end of the rubber tubing into a beaker that has been filled with limewater.

What is the purpose of this rubber tubing?

It allows carbon dioxide to escape, but prevents oxygen from entering.

Prepare the chart on the back of your Vee Form for recording the temperature of the solution of each vacuum bottle. Also record the time that the temperature was read. Allow for as many readings as possible in the next 48 hours. Record any other observations also.

Use the information on the **Knowing Side** of the Vee to interpret your results from the **Doing Side**, and then write your **Knowledge Claim**. Write a **Value Claim** for this study.

Analysis

Part A

1. In both set-ups, what do you think would happen if there were only one hole in the stopper for the thermometer?

The stopper would pop off due to pressure buildup.

2. What are some indications that a chemical reaction is taking place inside the vacuum bottles?

bubbles and rising temperature

3. What is the purpose of using a vacuum bottle, instead of another type container?

They are insulated to prevent heat loss.

The molasses energy graph should be a normal fermentation curve. If enough oxygen is dissolved in the sucrose solution and the temperature is high enough, cellular respiration may occur also. If a "glitch" at or near a certain threshold temperature appears, this could be indicative of some cellular respiration in addition to fermentation in the sucrose solution.

4. How does yeast release energy from sucrose?

by fermentation, and possibly cellular respiration

5. If you know that fermentation liberates energy and gives off carbon dioxide and alcohol as waste products, how would you prove that fermentation is really taking place in the sugar-yeast solution?

You could test for the presence of both carbon dioxide and alcohol.

6. What does your finished line graph indicate about this part of the experiment?

Temperature rises as fermentation takes place.

Part B

7. How did the observation of the rubber tubing in the beaker give evidence that a reaction was taking place?

A gas was produced.

8. What substance was produced?

carbon dioxide

9. What evidence was there that energy was being released?

The temperature increased.

10. Make a general conclusion about the energy-releasing process in living organisms.

Energy is released as heat; carbon dioxide is produced as a byproduct.

11. What was the source of the energy released in Part A? In Part B?

the sucrose; the molasses

Going Further

Generate a **New Focus Question** that could be the point of a new investigation based on what you now know.
• Design an experiment that would supply data about the source of energy for fermentation. Soon after the yeast is placed in molasses, take one drop of the solution and place it on a slide. Make observations of the yeast and estimate the number of cells seen with the microscope under high power. Repeat the above process on the second and third days. Relate the findings to the energy-releasing process.

Name _____ Date _____ Class _____

Release of Energy

KNOWING SIDE

DOING SIDE

Focus Question:

What is the evidence that yeast cells undergo fermentation? How does fermentation compare with respiration?

Subject Area:

Cell theory
Theory of fermentation
Theory of energy transformation from cell respiration

New Focus Question:

Could the evidence for cell respiration be acquired by measuring carbon dioxide exhaled in humans?

Concepts:

cell
heat
bonds

Vocabulary:

fermentation
respiration
sucrose

Concept Statements:

1. Cells are the basic unit of structure and function of all living things.
2. Sucrose contains energy in its bonds, that is released during cell respiration and fermentation. Sucrose is made by plants to store the sun's energy.
3. Energy can be released in the form of heat, which is easily recognized and measured.

Value Claim:

Organisms undergoing fermentation can be used to generate air bubbles in baked goods while the alcohol evaporates.

Knowledge Claim:

A. Large amounts of energy are released from the sucrose-yeast solution, as evidenced by the graph of solution temperature versus time.
B. Temperature rises as energy is released and carbon dioxide and alcohol are given off as waste products.

Records:

[See Additional Records and Observations on the following page].

Materials:

vacuum bottles 500 mL (2), two-hole stoppers for vacuum bottles (each with one thermometer and one 10-cm piece of glass tubing inserted), dried yeast package, sucrose 75 grams, molasses solution 750 mL, limewater 150 mL, flask 250 mL (1)

Procedure:

A. Observe the release of energy from a sucrose. B. Observe the process of fermentation. Record solution temperature over time in both A and B.

Table A Sucrose-Yeast Energy Production

Time Date Temperature	Time Date Temperature	Time Date Temperature
1.	8.	15.
2.	9.	16.
3.	10.	17.
4.	11.	18.
5.	12.	19.
6.	13.	20.
7.	14.	21.

Table B Molasses-Yeast Energy Production

Time Date Temperature	Time Date Temperature	Time Date Temperature
1.	8.	15.
2.	9.	16.
3.	10.	17.
4.	11.	18.
5.	12.	19.
6.	13.	20.
7.	14.	21.

Additional Records and Observations:

Graphs may vary. Typical fermentation curve shapes are shown.

Graph A Sucrose Energy Production

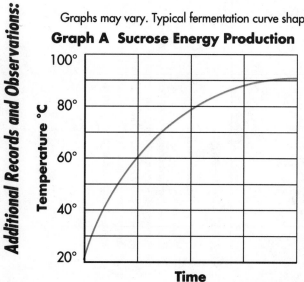

Graph B Molasses Energy Production

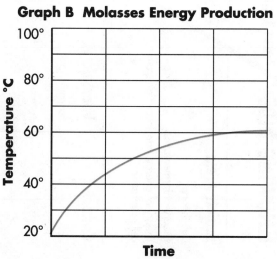

The Case of the Long-Lost Son

Introduction

This story is fictitious, but there are many examples in real life of situations similar to this one. Read the story and answer the questions that follow.

Mr. Cash died and left all his money to his two children. A young man claiming to be a lost third child sued for his share of the estate. The judge ordered blood tests for all family members and for the young man. Mr. Cash's blood type was AB. His wife had type A.

Materials

• Paper
• Pencil
• Textbook

Have students review Punnett squares in the text before they begin this investigation. Tell them that type O blood is recessive to both type A and type B blood.

Focus Question

Design your own **Focus Question** based on the information presented in the Introduction and Materials sections above, as well as the additional information presented in the Procedure section that follows. Write your Focus Question on the Vee Form.

Knowing Side

With the help of your teacher, your class will discuss the **Subject Area** background for this investigation. Then, with the help of your classmates, list the **Concepts** and the new **Vocabulary** words on your Vee Form. Now use these words in sentences that define and explain them in the Concept Statements section of your Vee Form.

Procedure

1. Use the following information to describe the possible genotypes and phenotypes of Mr. and Mrs. Cash's offspring:

   • The genotypes that produce phenotype A are AA and AO.
   • The only genotype that produces phenotype AB is AB.
   • The only genotype that produces phenotype O is OO.

2. Using the Punnett square in the **Records** section of the Vee form, diagram the offspring that could be produced if Mrs. Cash had genotype AA and Mr. Cash had genotype AB.

Which phenotypes could result among the offspring of this marriage?

Type A and Type AB

3. Using the Punnett square in the Records section of the Vee form, diagram the possible offspring if Mrs. Cash had genotype AO and Mr. Cash had genotype AB.

Which phenotypes could result among the offspring of this marriage?

Type A, Type AB, Type B

4. The man claiming to be the long-lost son then went for his blood test. He had type O blood.

Analysis

1. What are the possible genotypes of a person having blood type O?

only OO

2. What is the genotype of the young man claiming to be Mr. Cash's long-lost son?

OO

Going Further

Write a **New Focus Question** on your Vee Form that could be the point of a new investigation based on what you have learned.

The Case of the Long-Lost Son

KNOWING SIDE

DOING SIDE

Focus Question:

Is the man claiming to be the long-lost son of Mr. Cash truly his heir?

Subject Area:

Theory of ABO blood group inheritance

Value Claim:

Knowing more about the inheritance of certain traits enables the prediction of offspring in traits.

Knowledge Claim:

Since the parents could not produce an OO child, because Mr. Cash had no O to pass on, the claimant is either an impostor, or an illegitimate son of Mrs. Cash who was passed off as a son of Mr. Cash.

New Focus Question:

If a blood typing test is inconclusive, are there other genetic tests that may establish or disprove claims regarding the parentage of an individual?

Concepts:

Punnett square inheritance

ABO blood typing system

Vocabulary:

genotype
phenotype

Records:

Punnett Squares

	A	B
A	AA	AB
A	AA	AB

	A	B
A	AA	AB
O	AO	BO

Materials:

pencil, paper, textbook

Procedure:

Using the given parental blood genotypes, predict the possible blood genotypes and phenotypes of their offspring.

Concept Statements:

1. Blood types are inherited.
2. Punnett squares are used to predict possible genetic combinations of offspring.
3. Genotype is what the alleles actually are.
4. Phenotype is the expression of alleles.
5. A and B are dominant to O.
6. AA, AO, BB, BO, OO, and AB are the only possible ABO blood types.

Additional Records and Observations:

A Human Pedigree

Introduction

A sex-linked characteristic is determined by an allele that is carried only on the X chromosome. The shorter Y chromosome does not carry an allele for a sex-linked trait. Most sex-linked traits are recessive. Since there is only one X chromosome in the male, a male who carries a particular recessive allele on the X chromosome will have the sex-linked condition. A female who carries a recessive allele in one X chromosome will not have the condition if there is a dominant allele on her other X chromosome. She will express the recessive condition only if she inherits two recessive alleles—one from each parent. Her chances of inheriting a sex-linked condition are thus significantly less than those of a male.

Materials

• Pencil
• Paper

Focus Question

Design your own **Focus Question** based on the information presented in the Introduction and Materials sections above, as well as the additional information presented in the Procedure section that follows. Write your Focus Question on the Vee Form.

Knowing Side

With the help of your teacher, your class will discuss the **Subject Area** background for this investigation. Then, with the help of your classmates, list the **Concepts** and the new **Vocabulary** words on your Vee Form. In the **Concept Statements** section of your Vee Form, use these words in sentences that define and explain them.

Procedure

You may wish to go through the Procedure steps with the students for the first pedigree (Figure 1). Let them analyze the other pedigree (Figure 2) on their own.

1. Study the pedigree for hemophilia shown in Figure 1 below. In a pedigree, a square represents a male. If it is darkened, he has hemophilia; if clear, he has normal blood clotting.

Figure 1 Pedigree of heredity

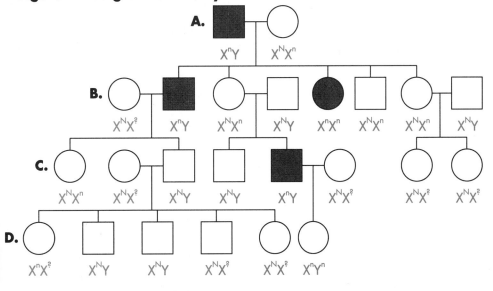

How many males represented by this pedigree have hemophilia? How many males are normal?

Three have hemophilia; eight are normal.

2. A circle represents a female. If it is darkened, she has hemophilia; if clear, she is normal.

How many females in this pedigree have hemophilia? How many are normal?

One has hemophilia; twelve are normal.

3. A marriage is indicated by a horizontal line connecting a circle to a square.

How many marriages are indicated in the pedigree?

six

4. A line perpendicular to a marriage line indicates the offspring. If the line ends with either a circle or square, the couple had only one child. However, if the line is connected to another horizontal line, then several children were produced, each indicated by a short vertical line connected to the horizontal line. The first child appears to the left and the last child to the right.

How many children did the first couple have?

five

5. Generation B represents the second generation. Generation C represents the third generation, and so on.

How many generations are represented in this pedigree?

four

6. The genotypes of the males in a pedigree for hemophilia are easy to determine, since normal blood clotting (N) is dominant and hemophilia is recessive (n). Since these alleles are on the X chromosome only, a male represented by a clear square will have the genotype denoted by X^NY. One represented by a darkened square will be X^nY. Label the genotypes of all the squares in the pedigree on your Vee form.

How many males have the genotype X^NY?

eight

7. Females who have hemophilia have an easy genotype to identify. They are $X^n X^n$. Both recessive alleles must be present for a female to have hemophilia. If one dominant allele (X^N) is present, the individual must be normal for clotting. Label all the females with hemophilia as genotype $X^n X^n$ on the pedigree.

How many women have genotype $X^n X^n$ in this pedigree?

one

8. Females who do not show the trait for hemophilia may be homozygous dominant ($X^N X^N$) or heterozygous ($X^N X^n$). A heterozygous female is called a carrier. Examination of traits among the offspring can often determine which genotype the parents have. If any child (son or daughter) has hemophilia, then the female must be heterozygous ($X^N X^n$). If her son has hemophilia, he has genotype $X^n Y$. He inherited the Y from the father, so the other allele in his genotype (X^n) had to come from the mother. If a daughter has hemophilia ($X^n X^n$), she inherited an X^n from each parent, thus making the genotype of the normal mother $X^N X^n$. Label all females $X^N X^n$ who have children with hemophilia.

What is the genotype of the female in the first generation?

$X^N X^n$

9. A female who has more than four sons, with none exhibiting hemophilia, is likely to have a genotype $X^N X^N$. If she has had four or fewer sons, her genotype is less certain. In such cases, her genotype is labeled as $X^N X^?$. Label the rest of the females in the pedigree as $X^N X^?$ or $X^N X^N$.

How many females in the pedigree have the genotype $X^N X^N$?

10. Label the genotype of all people represented in the pedigree for Figure 1.
11. Label the genotype of the individuals in the second pedigree (Figure 2) in the **Additional Records and Observations** section on the back of your Vee Form.
12. Fill in the table on the back of your Vee Form. It is based on the pedigree in Figure 2.

Use the information on the **Knowing Side** of the Vee to interpret your results from the **Doing Side**, and then write your **Knowledge Claim**. Write a **Value Claim** for this study.

Analysis

1. **Which sex usually inherits a sex-linked condition? Explain.**

   Male. The male has only one X chromosome. Any recessive condition on that X, since it is not countered by the Y, will be demonstrated.

2. How can you tell whether a female has a genotype X^NX^N, X^NX^n, or X^NX?

Examine the genotype of her children. For example, if any child has hemophilia, she is a carrier (X^NX^n). If she has more than four sons, and none exhibits hemophilia, then she is probably X^NX^N. If she has fewer than four sons, her genotype cannot be determined as reliably and is written X^NX?.

3. How many genotypes are possible in a pedigree of sex-linked traits? What are they?

six: X^NY, X^nY, X^NX^N, X^NX^n, X^NX?, X^nX^n

Going Further

In the appropriate space on your Vee Form, write a **New Focus Question** that could be the point of a new investigation based on what you have just learned.

• Look in the library for a pedigree of Queen Victoria and her descendants showing the incidence of hemophilia. Copy the pedigree and determine the genotypes of the members of this royal family.

Name _____ Date _____ Class _____

A Human Pedigree

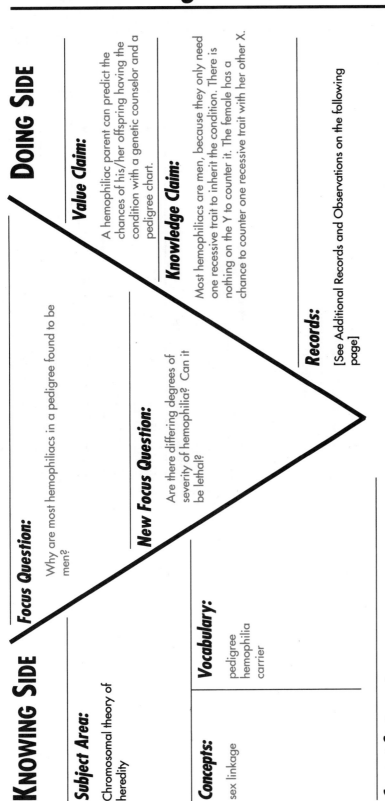

KNOWING SIDE

Focus Question:

Why are most hemophiliacs in a pedigree found to be men?

Subject Area:

Chromosomal theory of heredity

Concepts:

sex linkage

Vocabulary:

pedigree
hemophilia
carrier

Concept Statements:

1. A sex-linked trait is determined by an allele carried only on X chromosomes. The Y has no influence on a sex-linked trait.

2. Hemophilia is sex-linked and is a condition in which a blood clotting factor is missing.

3. A pedigree uses squares for males, circles for females, and darkened symbols to indicate presence of the trait in question.

4. A carrier possesses both the recessive trait and the dominant trait that masks it.

DOING SIDE

Value Claim:

A hemophiliac parent can predict the chances of his/her offspring having the condition with a genetic counselor and a pedigree chart.

Knowledge Claim:

Most hemophiliacs are men, because they only need one recessive trait to inherit the condition. There is nothing on the Y to counter it. The female has a chance to counter one recessive trait with her other X.

Records:

[See Additional Records and Observations on the following page]

New Focus Question:

Are there differing degrees of severity of hemophilia? Can it be lethal?

Materials:

pencil and paper

Procedure:

Complete two pedigrees, one from the marriage of a hemophiliac father and a carrier mother (Figure 1), the other from the marriage of a normal father and a hemophiliac mother (Figure 2).

Additional Records and Observations:

Figure 2

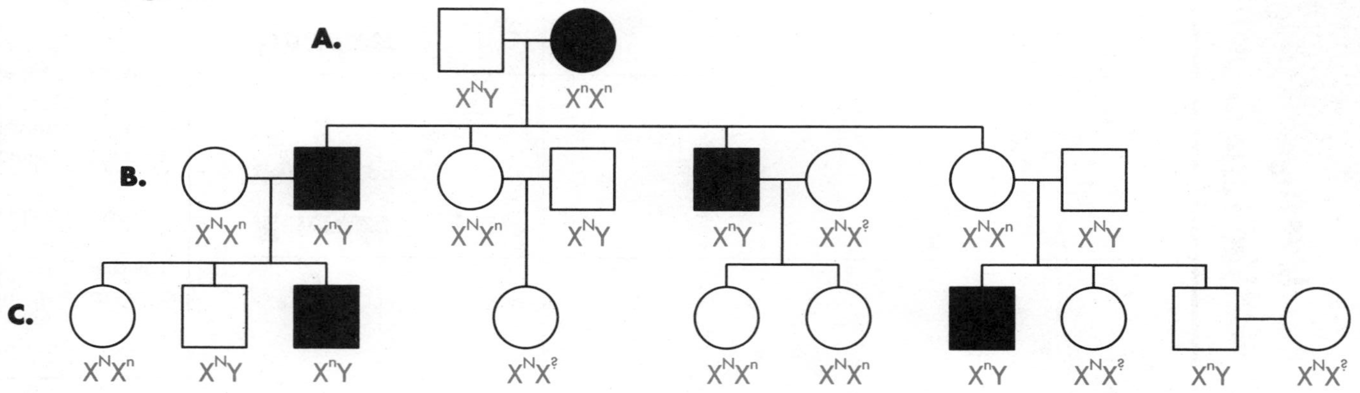

Questions About the Pedigree in Figure 2	Answer
Number of generations	3
Number of men with hemophilia	4
Number of women with normal blood clotting	10
Number of marriages	6
Number of men with genotype $X^N Y$	5
Number of women with genotype $X^n Y^n$	1
Number of single women	5
Number of people that never married	8
Number of women with genotype $X^n Y^2$	4
Number of couples with only one child	1

Protein Synthesis Drama

Introduction

Sometimes it is hard to visualize those biological processes that occur at the molecular level. Protein synthesis is one such process, controlled by DNA, which directs the manufacture of proteins, producing expressed traits. This lab provides a fun way of making these abstract ideas of molecular biology more concrete.

Materials

- Paper grocery bags (2)
- Scissors
- Markers (2 of different colors)
- Index cards (36), 3 in. x 5 in.
- Paper clips or safety pins (18)
- Dispenser of cellophane tape (1 per class)

Focus Question

Design your own **Focus Question** based on the information presented in the Introduction and Materials sections above, as well as the additional information presented in the Procedure section that follows. Write your Focus Question on the Vee Form.

Knowing Side

With the help of your teacher, your class will discuss the **Subject Area** background for this investigation. Then, with the help of your classmates, list the **Concepts** and the new **Vocabulary** words on your Vee Form. In the **Concept Statements** section of your Vee Form, use these words in sentences that define and explain them.

Procedure

1. Imagine that the classroom is a cell and the teacher's desk is the nucleus. Your teacher has a paper belt with DNA base letters and will play the part of the DNA molecule. Your teacher cannot leave the nucleus, and must produce a messenger RNA (mRNA) molecule that is able to leave the nucleus and carry the DNA instructions out into the cytoplasm to a ribosome.
2. Select a student to play the role of a ribosome. That person will remain by the chalkboard and will build a protein molecule by attaching amino acids end to end on the board later on in this drama. This role is very important, because the ribosome builds the protein by reading and following the DNA instructions written in the mRNA molecule. As the ribosome reads the mRNA molecule triplet codons one at a time, a transfer RNA (tRNA) molecule will deliver to the ribosome an amino acid specific to each codon. Throughout the drama, each player should refer to Table 1 on the following page in order to carry out the DNA instructions.
3. Several students are needed to play the tRNA molecules. Each of them receives an index card with an anticodon (a triplet of bases complementary to the mRNA codon) printed on it. They are to pin them onto their shirts. Each tRNA player also gets to hold the appropriate amino acid card (or "stop" card) to match his/her triplet codon.
4. The remaining students, if any, should try to follow the drama and point out where mistakes in the process, or mutations, could happen.

Table 1. The DNA, mRNA, and tRNA codons with corresponding amino acids.

These are in the form: DNA triplet-mRNA codon-tRNA anticodon-amino acid

AAA -UUU -AAA -Phenylalanine
AAC -UUG -AAC -Leucine
AAG -UUC -AAG -Phenylalanine
AAT -UUA -AAU -Leucine

ACA -UGU -ACA -Cysteine
ACC -UGG -ACC -Tryptophan
ACG -UGC -ACG -Cysteine
ACT -UGA -ACU -STOP

AGA -UCU -AGA -Serine
AGC -UCG -AGC -Serine
AGG -UCC -AGG -Serine
AGT -UCA -AGU -Serine

ATA -UAU -AUA -Tyrosine
ATC -UAG -AUC -STOP
ATG -UAC -AUG -Tyrosine
ATT -UAA -AUU -STOP

CAA -GUU -CAA -Valine
CAC -GUG -CAC -Valine
CAG -GUC -CAG -Valine
CAT -GUA -CAU -Valine

CCA -GGU -CCA -Glycine
CCC -GGG -CCC -Glycine
CCG -GGC -CCG -Glycine
CCT -GGA -CCU -Glycine

CGA -GCU -CGA -Alanine
CGC -GCG -CGC -Alanine
CGG -GCC -CGG -Alanine
CGT -GCA -CGU -Alanine

CTA -GAU -CUA -Aspartic acid
CTC -GAG -CUC -Glutamic acid
CTG -GAC -CUG -Aspartic acid
CTT -GAA -CUU -Glutamic acid

GAA -CUU -GAA -Leucine
GAC -CUG -GAC -Leucine
GAG -CUC -GAG -Leucine
GAT -CUA -GAU -Leucine

GCA -CGU -GCA -Arginine
GCC -CGG -GCC -Arginine
GCG -CGC -GCG -Arginine
GCT -CGA -GCU -Arginine

GGA -CCU -GGA -Proline
GGC -CCG -GGC -Proline
GGG -CCC -GGG -Proline
GGT -CCA -GGU -Proline

GTA -CAU -GUA -Histidine
GTC -CAG -GUC -Glutamine
GTG -CAC -GUG -Histidine
GTT -CAA -GUU -Glutamine

TAA -AUU -UAA -Isoleucine
TAC -AUG -UAC -Methionine
TAG -AUC -UAG -Isoleucine
TAT -AUA -UAU -Isoleucine

TCA -AGU -UCA -Serine
TCC -AGG -UCC -Arginine
TCG -AGC -UCG -Serine
TCT -AGA -UCU -Arginine

TGA -ACU -UGA -Threonine
TGC -ACG -UGC -Threonine
TGG -ACC -UGG -Threonine
TGT -ACA -UGU -Threonine

TTA -AAU -UUA -Asparagine
TTC -AAG -UUC -Lysine
TTG -AAC -UUG -Asparagine
TTT -AAA -UUU -Lysine

5. The ribosome has a job to do—make a protein! But even though the ribosome player is surrounded by amino acids, the proper ordering of these is unknown to him/her. Any given protein has a very specific arrangement of amino acids.

6. The teacher (DNA) is the only one with the instructions on how to build the protein, but cannot leave the nucleus to tell the ribosome what to do.

What can be done?

A messenger RNA molecule can be made.

At this point, pull out the strip of mRNA and hand this to the ribosome.

7. The ribosome will now read aloud the codons for making the protein one by one. Each tRNA player should check his/her anticodon and listen for the complementary codon, then deliver his/her amino acid to the ribosome, at the appropriate time.

8. The ribosome, as well as any members of the audience, should check the anticodon-codon match to be sure the tRNA players are correct.

9. The ribosome should tape the correct amino acid sequence to the chalkboard, amino acid after amino acid, until the "stop" codon is reached. The board now holds a representation of the finished product, a synthesized protein.

10. List the order of amino acids that your class placed on the board to make a protein in the **Records** section of your Vee Form.

Use the information on the **Knowing Side** of your Vee Form to interpret your results from the **Doing Side**, and then write your **Knowledge Claim** in the space provided. Write a **Value Claim** for this study.

Analysis

1. What is the role of DNA in the cell?

   DNA encodes the instructions for protein synthesis.

2. What is the role of mRNA?

   The mRNA copies those instructions (transcription) and delivers them to the ribosomes.

3. What is the role of tRNA?

   tRNA brings amino acids to the ribosome to match the mRNA codons and thus translates the DNA code into a protein chain.

4. What is the role of ribosomes?

   Ribosomes bring together the mRNA and the tRNA carrying amino acids, thus joining amino acids to form protein chains.

5. What is transcription?

   Transcription is the copying of the DNA code into mRNA codons.

6. What is translation?

   Translation is the joining of amino acids to form a protein chain, according to the sequence specified by mRNA codons.

7. How is this classroom drama different from what occurs in the cell?

There was no cytoplasm, nuclear membranes, etc.

8. Where could mutations have happened in the process?

Mistakes could have happened either during transcription or translation.

Going Further

In the appropriate space on your Vee Form, write a **New Focus Question** that could be the point of a new investigation based on what you have just learned.

In preparation for this activity:
1. Cut two 8 in. x 36 in. pieces of paper bag and tape them together in a 71-72 in. long loop that you can wear as a loose belt. Use a dark marker to print on the belt the following sequences of nucleotide bases:
    TAC–TTT–ACC–TAA–TGA–AGA–CGA–CTA–GTT–AAT–CCA–TTA–TCT–CTT–ACA–GTA–GGA–TTT–ATT
    on the top row, and the complementary base letters below for the second row:
    ATG–AAA–TGG–ATT–ACT–TCT–GCT–GAT–CAA–TTA–GGT–AAT–AGA–GAA–TGT–CAT–CCT–AAA–TAA
2. From the second paper bag, build a 6 inch wide x 6 feet long piece of paper. With a colored marker, print a single stranded sequence of As, Cs, Gs, and Us that are complementary to the first strand of the DNA model. This will be a model of mRNA.
    AUG–AAA–UGG–AUU–ACU–UCU–GCU–GAU–CAA–UUA–GGU–AAU–AGA–GAA–UGU–CAU–CCU–AAA–UAA
3. Print the name of the amino acid coded by each mRNA codon on separate index cards. Put a loop of tape on the back of the amino acid cards for attachment to the chalkboard during class.
    Figure 1 shows the amino acids that each mRNA codon codes for. The following is the correct order of amino acids for the first strand of DNA shown above:
    Methionine–lysine–tryptophan–isoleucine–threonine–serine–alanine–aspartic acid–glutamine–leucine–glycine–asparagine–arginine–glutamic acid–cysteine–histidine–proline–lysine–"stop"
4. On separate index cards, print the tRNA anticodons corresponding to each amino acid in the sequence. For example, for the codon GAU, which codes for the amino acid aspartic acid, the anticodon is CUA. Attach a paper clip (or safety pin) to each anticodon card. The students will wear these as they play the role of tRNA molecules.

Name ______________________ Date __________ Class __________

Protein Synthesis Drama

Investigation 7-1
Vee Form Report Sheet

KNOWING SIDE

Focus Question:

How does protein synthesis work?

New Focus Question:

What could cause mutations in this process?

Subject Area:

Protein synthesis

Concepts:

DNA
RNA

Vocabulary:

codon
anticodon
transcription
translation

Concept Statements:

1. Protein synthesis involves DNA, RNA, and amino acids. The proteins are assembled on the ribosomes.
2. DNA doesn't leave the nucleus.
3. A codon is a series of three bases that are matched by three complementary bases in its anticodon.
4. Protein synthesis involves transcription and translation. Transcription copies nuclear DNA code onto mRNA. During translation tRNA anticodons deliver amino acids in the correct sequence matching mRNA codons.

DOING SIDE

Value Claim:

Knowing how protein is made may enable genetic engineers to fix defective proteins someday.

Knowledge Claim:

DNA delivers its protein-making instructions to the ribosome via mRNA. The ribosome matches the triplet codons of mRNA to the anti-codons of tRNA, and their corresponding amino acids then are sequenced into a protein chain.

Records:

methionine-lysine-tryptophan-isoleucine-threonine-serine-alanine-aspartic acid-glutamine-leucine-glycine-asparagine-arginine-glutamic acid-cysteine-histidine-proline-lysine-STOP

Materials:

paper bags (2), scissors, two colored markers, 3-by-5 index cards (36), paper clips (18), dispenser of cellophane tape.

Procedure:

Put together a protein using people to play the roles of DNA, RNA, amino acids, and ribosomes.

Additional Records and Observations:

Investigating Human Karyotypes

Introduction

An excellent supplement to this lab that outlines genetic disorders is available from Ward's.

Landmarks of the Human Genome w/Reference Disk
(74 M 3673) Mac
(74 M 3674) IBM

Humans have 46 chromosomes in every body (somatic) cell. The reproductive cells, or gametes, have half this number. The chromosomes of the body cells can be arranged in homologous pairs according to their length and the position of their centromeres. Both males and females have 22 matched pairs of chromosomes. The twenty-third pair is matched in the female and is composed of two X chromosomes. The male, however, has an unmatched pair consisting of one X chromosome and a much shorter Y chromosome. The sex of an individual can be determined by making a karyotype from pictures of the chromosomes. If there are exactly 23 matched homologous pairs, the individual is a female. If there are 22 matched pairs and one unmatched pair, then the individual is a male.

Mistakes during meiosis resulting in an incorrect number of chromosomes in a gamete can lead to the offspring having a syndrome with specific abnormalities.

Materials

• Scissors
• Paper
• Tape

Point out to students that homologous chromosomes can be bent in different ways, but the length and centromere position are the same for a given pair.

Focus Question

Design your own **Focus Question** based on the information presented in the Introduction and Materials sections above, as well as the additional information presented in the Procedure section that follows. Write your Focus Question on the Vee Form.

Knowing Side

With the help of your teacher, your class will discuss the **Subject Area** background for this investigation. Then, with the help of your classmates, list the **Concepts** and the new **Vocabulary** words on your Vee Form. In the **Concept Statements** section of your Vee Form, use these words in sentences that define and explain them.

Procedure

1. Carefully cut rectangles around each of the chromosomes in Figure A in the **Additional Records and Observations** section on the back of your Vee Form, leaving a slight margin around each chromosome.
2. Arrange the chromosomes in pairs. The members of each pair will be the same length and will have the centromere in the same geometric location. Arrange the pairs according to their length.
3. Tape each pair to a piece of paper. Label each pair by making the longest pair #1, the next longest #2, and so on. This is Karyotype A.

Are there exactly 23 homologous pairs present?

No; there are 22 homologous pairs and 1 unmatched pair.

Attach the finished karyotype to your Vee Form. Use the information on the **Knowing Side** of the Vee to interpret your results from the **Doing Side**, and then determine your **Knowledge Claim**. Write a **Value Claim** for this study.

Analysis

1. How can homologous pairs be identified?

Both chromosomes have the same length and the same centromere location.

2. What is the major chromosomal difference between gametes and somatic cells?

Gametes have only one member of each homologous pair; somatic cells have both members of each homologous pair.

3. Based on the evidence obtained from Karyotype A, what is the sex of this individual? How do you know?

Male; since the members of the 23rd pair of chromosomes do not match

4. "The male determines the sex of the offspring." Explain this statement.

Sex is determined by which sex chromosome is supplied by the male. The female always contributes an X chromosome; the male may contribute an X or a Y.

Going Further

Write a **New Focus Question** on your Vee Form that could be the point of a new investigation based on what you have learned.
- If time and available materials permit, your teacher will provide you (or your lab group) with an unknown karyotype associated with a specific genetic disorder. Determine the karyotype and the disorder that accompanies it.

You may wish to obtain the following items from Ward's.

Human Chromosome Spreads:
(33 M 1048) Trisomy 21, Down Syndrome M, 47 XY+21
(33 M 1049) Trisomy 21, Down Syndrome F, 47 XX+21
(33 M 1050) Turner's Syndrome, 45 XO
(33 M 1050) Kleinfelter's Syndrome, 47 XXY

Human Karyotype Form:
33 M 1045

Give the students the following information about these genetic syndromes.
- **Trisomy 21:** The 21st pair of chromosomes has an extra copy. It is also called Down syndrome. These people show a range of mental retardation.
- **Turner's:** missing an X or a Y, resulting in female with genotype XO, who is sterile but without mental retardation.
- **Kleinfelter's:** XXY. male with an extra X chromosome who is sterile and may have below normal intelligence.

Give each student (or lab group) an unknown karyotype and have them determine the genetic syndrome associated with it.

Investigation 8-1
Investigating Human Karyotypes Vee Form Report Sheet

DOING SIDE

Value Claim:

Genetic counseling relies heavily on karyotyping.

Knowledge Claim:

Twenty-two homologous pairs can be identified in a karyotype. The sex of an individual can be determined from pair #23.

Records:

[See Additional Records and Observations on the following page]

Materials:

scissors, paper, tape

Procedure:

Cut out chromosomes from Figure A and arrange them in homologous pairs on a separate paper. Tape them down and look for sex determination as well as any abnormalities in number.

KNOWING SIDE

Focus Question:

What can be learned from preparing and analyzing human karyotypes?

New Focus Question:

How can different genes be recognized within these chromosomes?

Subject Area:

Chromosomal theory of heredity

Concepts:

chromosomes
homologous pairs
sex chromosomes
sex determination

Vocabulary:

karyotype
syndrome
somatic
gamete
centromere

Concept Statements:

1. Humans have 22 pairs of homologous somatic chromosomes. Females have 1 pair of homologous sex chromosomes, but males have an unmatched pair of sex chromosomes.
2. A karyotype is a chromosomal map of a cell's chromosome pairs.
3. Many genetic syndromes result from an incorrect number of chromosomes in a gamete.

Figure A

Transforming Genetic Information

Introduction

A *Euglena* culture kit that includes Detain™ is available from Ward's.
Euglena Culturing Kit
(87 M 0101)

Materials

Euglena is a unique organism because it is both autotrophic and heterotrophic. It contains chloroplasts and can perform photosynthesis, yet it must consume vitamins B_1 and B_{12} to survive. When grown in the dark, *Euglena* loses its chlorophyll and becomes a colorless consumer. When grown in the light, *Euglena* becomes green. In this investigation you will try to produce a mutant form or new variety of *Euglena*.

- Medicine droppers (3)
- *Euglena* culture
- Detain™ (Ward's) (protozoa slowing agent)
- Compound light microscope
- 100-mL beakers (3)
- Wax pencil
- Pond water
- 10-mL graduated cylinder
- incubator
- toothpicks

Focus Question

Design your own **Focus Question** based on the information presented in the Introduction and Materials sections above, as well as the additional information presented in the Procedure section that follows. Write your Focus Question on the Vee Form.

Knowing Side

With the help of your teacher, your class will discuss the **Subject Area** background for this investigation. Then, with the help of your classmates, list the **Concepts** and the new **Vocabulary** words on your Vee Form. In the **Concept Statements** section of your Vee Form, use these words in sentences that define and explain them.

Part A: *Normal* Euglena *Appearance*

Procedure

The pond water supplies nutrients for *Euglena* but must not contain it. Boil it and then filter with Whatman #1 filter paper. Alternatively, you may purchase **Euglena Media** from Ward's (88 M 5200)

Detain™ is also available separately. (37 M 7950)

1. Using a medicine dropper, place a drop of *Euglena* culture on a glass slide. Add a drop of Detain™ solution (to slow the organism down for better viewing), mix with a toothpick and add a coverslip.

Why is the solution green in color?

The *Euglena* contains chlorophyll.

2. Observe the slide at low power. Then switch to high power and focus on a single organism.
3. Make a drawing of the *Euglena* in the **Records** section of your Vee Form, and label the nucleus, cell membrane, chloroplasts, and flagellum.

Part B: *Experimental Setup*

1. With a wax pencil, put your initials on each of 3 beakers and label them 1, 2, and 3, respectively.
2. Add 5 mL of *Euglena* culture to each beaker. Then add enough fresh pond water to each beaker to fill each two-thirds full.

3. Place the beakers as indicated below:
   Beaker 1: in a cabinet or drawer where no light will be present but where normal room temperature is maintained.
   Beaker 2: near the window at room temperature where it can receive normal sunlight in the daytime.
   Beaker 3: in an incubator at 35°C with no outside light.
4. Allow beakers to remain undisturbed for one week.

Part C: Observations After One Week

1. Label three slides 1, 2, and 3.
2. Make a wet mount with Detain™, using the liquid from your beakers. Use a clean dropper for each wet mount.
3. Examine each slide under the microscope and record your observations in Table A in the **Additional Records and Observations** section on the back of your Vee Form.

Part D: Observations After Two Weeks

1. Add fresh pond water to all three beakers to fill them two-thirds full. Place the three beakers near a window. Allow them to remain undisturbed for one week.
2. At the end of the week, make a wet mount of *Euglena* from each beaker and observe each under a microscope.
3. Record your observations in Table A in the **Additional Records and Observations** section on the back of your Vee Form.

Use the information on the **Knowing Side** of the Vee to interpret your results from the **Doing Side**, and then write your **Knowledge Claim**. Write a **Value Claim** for this study.

Remind students to clean up their materials and wash their hands following Parts B, C, and D.

Analysis

1. What environment(s) altered the appearance of *Euglena* after one week? How was the appearance altered?

   Beakers 1 (grown in the dark) and 3 (grown in the incubator at 35°C) had altered

   appearances. No chloroplasts were present after one week. *Euglena* appeared bleached.

2. After the second week, when all three beakers had been exposed to light, which beakers contained *Euglena* that were most similar? How were they similar?

   Beakers 1 (grown in the light) and 2 (grown in the dark) both contained *Euglena* with

   chloroplasts.

3. Did any chloroplasts appear in *Euglena* cultured at 35°C after they were placed back at normal room temperature and in light? Why?

   No. Something in the *Euglena* metabolism was altered that created the new colorless

   consumer variety.

Going Further

Generate a **New Focus Question** which could be the point of a new investigation based on what you now know.
• Design an experiment to test whether *Euglena* without chloroplasts will pass this trait to their offspring.

Transforming Genetic Information

Investigation 8-2
Vee Form Report Sheet

KNOWING SIDE

DOING SIDE

Focus Question:

How can we transform the genetic material in *Euglena* to create a mutant strain?

Subject Area:

Theory of genetic mutation

New Focus Question:

Will the "bleached" *Euglena* pass this trait to their offspring? How does radiation affect *Euglena*?

Concepts:

heterotrophic
autotrophic
photosynthesis
consumer

Vocabulary:

Euglena
mutant
flagellum

Concept Statements:

1. *Euglena* is a chloroplast-containing organism that is photosynthetic in light and heterotrophic in the dark.
2. In darkness, *Euglena* loses chlorophyll and becomes colorless only to have its chlorophyll return in the light.
3. Mutants have part of their genetic material transformed from its normal condition to something abnormal.

Value Claim:

Environmental factors must be considered when mutations may lead to undesirable changes in organisms.

Knowledge Claim:

A mutant of *Euglena* that will permanently lack chlorophyll can be created in a week's time by altering the environmental factors light and temperature.

Records:

[See Additional Records and Observations on the following page.]

Materials:

3 medicine droppers, *Euglena* culture, Detain™, microscope, three 100-mL beakers, wax pencil, pond water, 10-mL graduated cylinder, incubator, toothpicks

Procedure:

Place beakers of *Euglena* culture in three different temperature and light environments for one week and observe. Then test for return to normalcy by placing all three in light at room temperature for one more week.

Additional Records and Observations:

Drawings of Euglena:

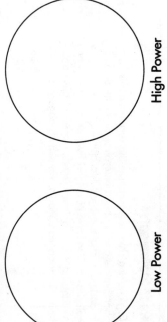

Low Power

High Power

Table A Euglena Appearance

Beaker/Environment	After 1 Week	After 2 Weeks (light/room temp)
#1/Dark	No chloroplast	Chloroplast present
#2/Light	Chloroplast present	Chloroplast present
#3/35°C	No chloroplast	No chloroplast

Peppered Moth Survey

Introduction

Industrial melanism is the term used to describe the adaptation of an organism in response to a type of industrial pollution. One example of rapid industrial melanism occurred in the peppered moth, *Biston betularia*, in the area of Manchester, England, from 1845 to 1890.

Before the Industrial Revolution, the trees in the forest around Manchester were light grayish-green due to the presence of lichens on their trunks. Peppered moths, which lived in the area, were colored light with dark spots. Their coloring served as protective camouflage against predators, especially birds. As the Industrial Revolution progressed, the trees became covered with soot, turning the trunks dark. Over a period of 45 years, a change in the peppered moths took place.

In this investigation, you will observe the effects of industrial melanism in the peppered moths over the course of 10 years. You will then determine the relationship between the environmental changes and the color variation of the peppered moth by using research data to graph the effects of an environmental adaptation.

Materials

• Graph paper
• Colored pencils (2) (optional)

Focus Question

Knowing Side

Procedure

Design your own **Focus Question** based on the information presented in the Introduction and Materials sections above, as well as the additional information presented in the Procedure section that follows. Write your Focus Question on the Vee Form.

With the help of your teacher, your class will discuss the **Subject Area** background for this investigation. Then, with the help of your classmates, list the **Concepts** and the new **Vocabulary** words on your Vee Form. In the **Concept Statements** section of your Vee Form, use these words in sentences that define and explain them.

1. Table A contains data from a 10-year study of two varieties of the same species of peppered moth. The numbers represent moths captured in each of 10 consecutive years. The traps were located in the same area each year.

Table A

Year	Number of Light Moths Captured	Number of Dark Moths Captured
2	537	112
3	484	198
4	392	210
5	246	281
6	225	357
7	193	412
8	147	503
9	84	594
10	56	638

2. Using the data provided in Table A, construct a graph in the **Additional Records and Observations** section of your Vee comparing the numbers of each variety of peppered moth. Label the axes with the years of the study (plotted horizontally) and the number of moths captured (plotted vertically). Use different colored pencils (or a solid line and a dotted line) to indicate each of the two color variations of the moth. Be sure to inclue in a key beneath the graph.

3. Use your graph and your textbook, if needed, to answer the following questions:

What preys on the peppered moths?

birds

Students need only construct one graph using two different markings to show the variations in the moth.

If the bark of trees is dark and the moths that rest there are light, what might happen to the moths?

They will be seen easily by birds, so they will likely be captured.

What is a mutation?

A mutation is a change in the normal genetic makeup of an individual that is sometimes expressed in the appearance of the offspring.

What could have caused the first few moths to change from a light variety to a dark variety?

random mutations

What event caused the tree trunks of many trees in England to turn from light to dark?

the Industrial Revolution released large amounts of soot into the atmosphere. This material settled on the tree trunks and caused them to darken.

Which variety of moth increased over the 10-year period?

the dark variety

What is the name of this type of evolutionary change?

industrial melanism

Use the information on the **Knowing Side** of the Vee to interpret your results from the **Doing Side**, and then to help you determine your **Knowledge Claim**. Write a **Value Claim** for this study.

Analysis

1. Using the data on the graph, draw a conclusion concerning the population of peppered moths in the sampled area of England.

   Over a period of 10 years, the number of light-colored peppered moths decreased, and the number of dark-colored moths increased.

2. Explain the reason for the increase in the number of dark-colored moths.

   The dark moths were capable of blending into the new environment. Therefore, they were more effectively hidden from predators, and so were better suited to survive.

3. What means could be used to return the environment of the peppered moth to its original state?

stringent pollution controls, alternate energy sources

4. What effect would cleaning up the environment have on the moths?

There would probably be a decrease in the number of dark-colored moths and an increase in the number of light-colored moths. This would be a reversal of the trends shown on the graph.

Going Further

In the appropriate space on your Vee Form, write a **New Focus Question** that could be the point of a new investigation based on what you have learned.

• Research another organism that has shown a dramatic adaptation over a relatively short period of time, such as pesticide-resistant insects or antibiotic-resistant bacteria. Prepare a report on the circumstances surrounding this event.

Name _____ Date _____ Class _____

Peppered Moth Survey

KNOWING SIDE

DOING SIDE

Focus Question:

How did the Industrial Revolution affect the population of peppered moths in the sampled area of England?

Subject Area:

Evolution
Natural selection

New Focus Question:

Since the cutback on soot from industry near Manchester, has the local peppered moth population returned to the original light color with dark spots?

Value Claim:

Knowing the effects of environmental change on a species is important to its survival. Ecological research must examine the effects of environmental pollution on organisms.

Knowledge Claim:

As the industrial soot covered the trees, the light moths were more easily seen by birds and their numbers decreased. The dark ones became more plentiful, as their coloration provided camouflage.

Records:

[See Additional Records and Observations on the following page]

Materials:

graph paper, colored pencils (2)

Procedure:

Produce a graph from a data table comparing numbers of two varieties of peppered moths.

Concepts:

industrial melanism
selection
survival
environmental
change

Vocabulary:

melanin
camouflage
adaptation
mutation

Concept Statements:

1. Those individuals who are best adapted to environmental changes are those naturally selected to survive and reproduce.

2. Industrial melanism describes the adaptation of an organism in response to industrial pollution, an environmental change.

3. Melanin is a pigment that gives peppered moths their darker wing color.

4. Camouflage is the blending of an individual with its surroundings.

5. An adaptation is a trait arising from a mutation already possessed by an individual. It becomes valuable to survival when the environment changes favorably for organisms possessing the trait.

Additional Records and Observations:

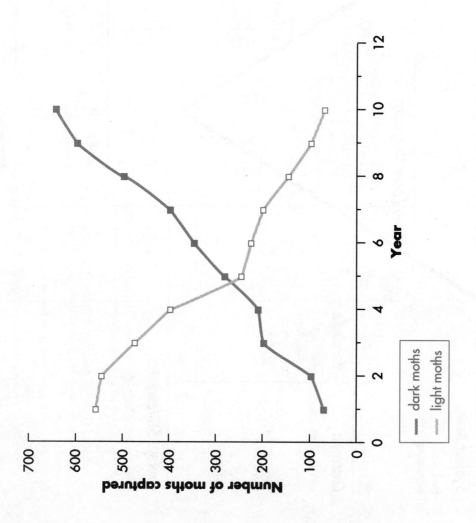

Number of moths captured

Year

dark moths
light moths

Animals of the Future

Introduction

You will see a number of drawings of "future animals" in this investigation. They were designed by a British fossil expert, Dougal Dixon. They are fantasy animals, but Dixon believes that they may be similar to real animals that might exist 50 million years in the future.

Materials

• Textbook

Focus Question

Design your own **Focus Question** based on the information presented in the Introduction and Materials sections above, as well as the additional information presented in the Procedure section that follows. Write your Focus Question on the Vee Form.

Knowing Side

With the help of your teacher, your class will discuss the **Subject Area** background for this investigation. Then, with the help of your classmates, list the **Concepts** and the new **Vocabulary** words on your Vee Form. In the **Concept Statements** section of your Vee Form, use these words in sentences that define and explain them.

Procedure

1. As you look at the animals pictured on the following pages, record the following information in the data table provided in the **Additional Records and Observations** section on the back of your Vee Form:
   a. the animal's body structure
   b. any special characteristic that would help the animal survive in its habitat
   c. any ideas as to who/what its ancestors were, its probable diet, and possible predators or prey
2. After observing the drawings carefully, answer the following questions. You may use your textbook to help you find the answers.

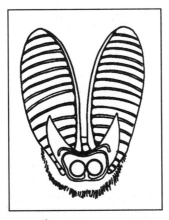

Purrip bat

Night stalker

What does predatory mean?

A predatory animal catches another live animal and consumes it for food.

Why do you suppose that the Purrip bat has such large ears?

Bats are nocturnal animals and have poor eyesight. They use ears like "radar dishes" to locate prey and to avoid hitting objects in the dark. Thus, larger-eared bats would have a survival advantage over smaller-eared bats.

Why has the Night stalker bat lost its wings?

It is about four feet tall, and its wings have evolved into clawed hind legs that reach over its shoulders. Being much larger than its ancestor, the flying bat, it is too heavy to fly. It must hunt much larger prey to satisfy its nutritional needs.

Flooer bat

Gigantelope

What adaptations does the Flooer bat have to compensate for its loss of ability to fly?

It may be an insectivore, and its large ears and nose may mimic flowers and attract insects.

What characteristics does the Gigantelope share with its antelope and elephant ancestors?

It has horns similar to an antelope, and is comparable in size to an elephant. It is a herbivore, like the antelope and elephant.

Zarander

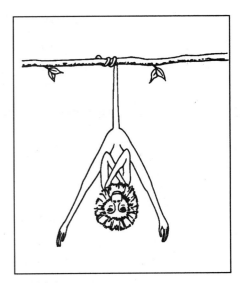

Ziddah

What is the Zarander descended from? Why is it so much larger than its ancestor?

Possibly a pig; answers will vary. The animal feeds exclusively on vegetation, and most

large animals are herbivores.

From what is the long-armed Ziddah probably descended?

Since it has a tail, it is probably descended from a monkey.

Name any other primate in the set of pictures.

There are no other primates.

Parashrew   Rootsucker

The Parashrew uses its parachute tail to glide from its nest. From what animal(s) do you think the Parashrew is descended?

possibly the shrew and the flying squirrel

Speculate on the probable diet of the Parashrew.

Insects; if it is descended from the shrew, which is an insectivore, it is probably an insectivore also.

What advantages do long pointed claws, a horny head shield, and a nutlike shell offer to the desert-dwelling Rootsucker?

The head and claws allow it to dig in the sand; the hard shell helps prevent water loss.

From what animal is the Rootsucker probably descended?

possibly a mole of some kind, or an armadillo

Truteal   Pfrit

The Truteal pictured is bright blue, with an insect-catching beak. It is sightless and has sensitive whiskers and highly developed hearing. From what animal is it probably descended?

A shrew.

What advantage could being blue offer this animal?

It might mimic something that is poisonous or tastes unpleasant. This could protect it from being eaten.

Describe the unusual feature of the Pfrit.

It has a long beak for catching insects.

Reedstilt

The Reedstilt hides among the reeds near a pond and catches passing fish. What characteristics does the Reedstilt possess that would contribute to its success?

The hair on its legs may look like food to passing fish. Its sharp teeth and long beak-like mouth are efficient for catching and holding fish. Its striped body covering provides good camouflage.

Chirit

What animal does the Chirit resemble?

the squirrel

Speculate on the diet and the habitat of the Chirit.

It eats greens, nuts, and fruits; trees are its probable habitat.

Now that you have a summary of the results, use the information on the **Knowing Side** of the Vee to interpret your results from the **Doing Side**, and then determine your **Knowledge Claim**. Write a **Value Claim** for this study.

Analysis

1. What are the most probable food sources of the animals presented in the investigation? How can this be accounted for?

Most are insect eaters and/or herbivores. Over 50 million years, the natural habitats

of many carnivores might be disrupted as a result of human activity.

Going Further

Generate a **New Focus Question** that could be the point of a new investigation based on what you now know.

Name _____ Date _____ Class _____

Animals of the Future

KNOWING SIDE

DOING SIDE

Subject Area:

Evolution
Fantasy

Focus Question:

How might animals evolve in the future in response to environmental change?

Value Claim:

Knowing how environmental changes may change future life forms could prevent humans from destroying their food sources.

New Focus Question:

Could we take a population of animals and their predators and prey and change a controlled captive environment so greatly as to force some changes in 50-100 years?

Knowledge Claim:

The adaptations of these imagination-created animals seem to be new specializations of animals we know today. However, their prey population or predator population must have changed.

Concepts:

evolution
environmental change

Vocabulary:

fossil
predator

Records:

[See Additional Records and Observations on the following page]

Materials:

textbook

Concept Statements:

1. On the basis of past evolutionary fossil data, fantasy animals like these might actually exist 50 million years in the future.
2. Predators are animals who eat other animals.
3. Fossils are evidence of life from a very long time ago.
4. Evolution is the change living organisms undergo over a long period of time in response to environmental change.

Procedure:

Examine the drawings of several fantasy animals and observe special characteristics that might help the animal survive in its habitat.

Additional Records and Observations:

Animal	Special Characteristics Observed	Speculations on Possible Diet, Possible Predators, etc.
Purrip bat		
Night stalker		
Flooer bat		
Gigantelope		
Zarander		
Ziddah		
Parashrew		
Rootsucker		
Truteal		
Pfrit		
Reedstilt		
Chirit		

Human Evolution

Introduction

Physical anthropologists study fossils to learn about human evolution. Often these scientists use fossil fragments to reconstruct a complete jaw or skull. From these few clues, a physical anthropologist must be able to distinguish and classify fossils of different species. One way these scientists classify fossils is by comparing certain anatomical structures of the fossils with those of other hominid fossils, modern apes, and modern humans. In this investigation, you will compare anatomical features of apes, hominids, and humans to determine which hominids are more ape-like and which are more humanlike.

Materials

• Metric ruler
• Protractor
• Calculator

Focus Question

Design your own **Focus Question** based on the information presented in the Introduction and Materials sections above, as well as the additional information presented in the Procedure section that follows. Write your Focus Question on the Vee Form.

Knowing Side

With the help of your teacher, your class will discuss the **Subject Area** background for this investigation. Then, with the help of your classmates, list the **Concepts** and the new **Vocabulary** words on your Vee Form. In the **Concept Statements** section of your Vee Form, use these words in sentences that define and explain them.

Procedure

Part A: Primate Hands

Humans, apes, and other primates have grasping hands with opposable thumbs. Opposable thumbs can touch the tip of each finger, enabling primates to easily manipulate objects and use a variety of tools. Unlike human hands, the hands of gorillas and other apes are used for both walking and grasping. The opposable thumbs in the gorilla and human are examples of homologous structures, which are body parts with structurally similar features. Study the drawings of the primate hands below. Compare the gorilla hand with the human hand.

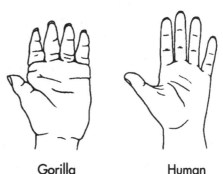

Gorilla   Human

In what ways do the human hand and the ape hand appear similar?

Both have five digits and an opposable thumb.

How are the human hand and the ape hand different?

Humans have long slender fingers, while the ape's fingers are shorter and thicker. Also, the human thumb is longer in proportion to the other fingers than is the ape thumb.

Part B: Ape and Human Skulls

Study the ape and human cranial cavities, faces, teeth, and jaws in the drawings below. Refer to these drawings as you make the observations and measurements described in questions 1-7 below. Record your results in the data table on the **Additional Records and Observations** portion on the back of your Vee Form.

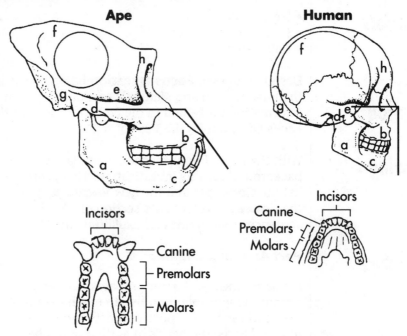

Due to measurement uncertainty, student answers may vary somewhat. However, they should be fairly close to those given in the data table.

1. The circle drawn on each skull represents the cranial capacity of each primate. Measure the radius of each circle in centimeters, then cube this number. Now multiply the result by 1,000 to approximate the cranial capacity size in cubic centimeters. A large cranial capacity is characteristic of modern humans. Record these measurements in the data table.

2. Measure the distance from a to b and from c to d in centimeters for each skull. Multiply these two numbers together, then multiply the product by 40. The result is the approximate size of the lower face area in square centimeters. Less face area is typical of modern humans. Record these measurements in the data table.

3. As in Step 2, measure e to f and g to h in centimeters for each skull. Multiply these two numbers together, then multiply the product by 40. The result is the approximate size of the brain area in square centimeters. More brain area is typical of modern humans. Record these measurements in the data table.

4. Note the two lines in the nose area of each skull. Measure the inside angle with your protractor to approximate the jaw angle. This measurement will indicate how far the jaw protrudes outward. A jaw angle of approximately 90° is a trait of modern humans. Record this information in the data table.

5. The brow ridges are the bony ridges above the eye sockets. In your table, note their presence or absence in each skull. Modern humans have lost the characteristic of having prominent brow ridges.

6. Note the number of teeth in the lower jaw, and the number of each kind of tooth. Record this information in the data table.

7. Compare the length of human canine teeth with the length of the ape canine teeth. In your table, designate the canine teeth as either "long" or "reduced."

Part C: Hominid Skulls

1. Study the four fossil hominids shown in the drawings below.

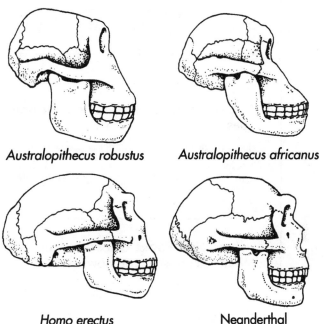

Australopithecus robustus   *Australopithecus africanus*

Homo erectus   Neanderthal

In what ways does each hominid skull resemble the ape skull in Part B?

All have prominent brow ridges and a broad lower jaw. Except Neanderthal, all have a protruding lower face.

How does each hominid skull resemble the human skull shown in Part B?

All have reduced canine teeth and a rounder braincase. The skulls of *Homo erectus* and Neanderthal have a large brain size, like the human skull.

2. Analyze each skull using the same techniques you applied when you compared ape and human skulls in Part B. Measure as many features as you can for each of the four skulls. Note the presence or absence of brow ridges in each skull. Determine whether the canine teeth are long or reduced. Record your data in the second data table on the back of your Vee Form.

3. For each hominid, classify each feature as being more apelike (A), more humanlike (H), or intermediate (I) by placing the appropriate letter in each column in the table on the back of the Vee Form.

Use the information on the **Knowing Side** of the Vee to interpret your results from the **Doing Side**, and then determine your **Knowledge Claim**. Write a **Value Claim** for this study.

Analysis

1. What do homologous structures such as the common opposable thumb suggest about humans and gorillas?

   They share a common ancestor.

2. Describe how differences between human hands and ape hands are related to their functions.

   The long, slender thumb and fingers give humans the increased dexterity required for manipulating tools. The short, full fingers of the ape enable it to achieve greater walking balance.

3. Compare the skull features and the brain size of humans with those of apes.

   Humans have a greater cranial capacity than apes. The lower face area of apes is much greater than that of humans. Apes also have large brow ridges, which are absent in humans. The jaw angle in apes is greater than that on humans.

Going Further

Generate a **New Focus Question** that could be the point of a new investigation based on what you now know.

If available, you may want to use life-size reproductions or casts of ape, hominid, and human skulls. If you have life-size casts of skulls available for students to measure, to calculate brain area, instead of multiplying by the factor of 40, simply multiply the distance between a and b by that between c and d . To calculate cranial volume, measure the radius of the sphere of the skull, cube this radius and then multiply by 4.2 (or $4/3 \times pi$).

Name _____ Date _____ Class _____

Human Evolution

KNOWING SIDE

DOING SIDE

Focus Question:
Are the human and ape closely related, based on hand and skull comparisons? Which hominids are more or less closely related, based on fossil evidence?

New Focus Question:
What does the DNA evidence suggest about fossil hominids and modern apes and humans?

Subject Area:
Evolution

Concepts:
homologous structures
cranial capacity

Vocabulary:
hominids
anthropologist

Concept Statements:
1. Physical anthropologists study fossils to learn about human evolution.
2. Comparisons of anatomical features of apes, hominids, and humans help determine which hominids are more apelike and which are more humanlike.
3. Homologous structures are those that are similar in anatomy, not necessarily in function.

Value Claim:
With more comparison of amino acids and DNA from fossils, it is likely that the future archaeologist may rely on skeletal reconstruction to a lesser extent.

Knowledge Claim:
Both humans and apes have five fingers and an opposable thumb, but human fingers are long and thin and the thumb is longer. Humans have a greater cranial capacity than apes. Intermediate fossils show both humanlike and apelike qualities.

Records:
[See Additional Records and Observations on the following page]

Materials:
metric ruler, protractor, calculator

Procedure:
Compare and contrast the features of fossil and modern ape and hominid hands, skulls, and teeth.

Part B

	Human	Ape
Cranial Capacity (cubic cm)	1950 cm^3 (approx.)	550 cm^3 (approx.)
Lower Face Area (square cm)	80 cm^2 (approx.)	300 cm^2 (approx.)
Brain Area (square cm)	250 cm^2 (approx.)	220 cm^2 (approx.)
Jaw Angle (degrees)	90°	130°
Brow Ridges (absent/present)	absent	present
Number of Teeth (lower jaw)	16	16
molars	6	6
premolars	4	4
canines	2	2
incisors	4	4
Canine Teeth (long/reduced)	reduced	long

Part C

	Australo-pithecus robustus	Australo-pithecus africanus	Homo erectus	Neanderthal
Lower Face Area (square cm)	A	A	I	H
Brain Area (square cm)	A	A	I	H
Jaw Angle (degrees)	A	A	I	H
Brow Ridges (absent/present)	present	present	present	present
Canine Teeth (long/reduced)	reduced	reduced	reduced	reduced

Additional Records and Observations:

Mapping an Environmental Site

Introduction

The environmental sites need to be carefully chosen. The sites can be very similar to simplify the investigation or they can be made more challenging by including such variables as sloping/level terrain, and shady/sunny environment.

Materials

Focus Question

Knowing Side

Procedure

This investigation should be done after the sections on the transfer of energy, trophic levels, food chains, and food webs have been covered in the text. You may want to divide the class into groups, with each group responsible for marking a particular color-coded biotic factor on the map.

Ecology is the study of the relationship between living things and their environments. In this investigation, you are going to play the role of an ecologist. You will observe and collect data about an environmental site, including the kinds of organisms and the relative abundance of the organisms inhabiting the site. These are the biotic factors. You will also need to observe and collect data about the physical features of the environment, or the abiotic factors. You will examine the interrelationships among these factors.

- Meter stick
- Stakes (4)
- String
- Poster board (square)
- Large felt-tip pen
- Tags (240, 30 of each of 8 colors)
- Glue

Design your own **Focus Question** based on the information presented in the Introduction and Materials sections above, as well as the additional information presented in the Procedure section that follows. Write your Focus Question on the Vee Form.

With the help of your teacher, your class will discuss the **Subject Area** background for this investigation. Then, with the help of your classmates, list the **Concepts** and the new **Vocabulary** words on your Vee Form. Now use these words in sentences that define and explain them in the **Concept Statements** section of your Vee Form.

1. With a meter stick, measure a 100 square meter site to be studied. Place one stake at each corner of the site. Loop string around one stake and continue to the next stake until boundaries have been formed for the site.

2. Draw the physical features of the site on the poster board. This will be the Records section of the **Doing Side** of your Vee, although it is too big to fit on the Vee form. Examples of physical features might be streams, sidewalks, or trails. Let the entire poster board represent the area of your site.

3. Add biotic factors to the map with the colored tags, using this key:
   medium green—tree
   red—weeds
   light green—bush
   yellow—grass
   brown—evidence of animals (burrows, nests, egg cases)
   blue—animals (ants, grasshoppers, birds)
   black—dead materials (logs or dead animals or trees)
   orange—fungi (mushrooms, etc.)
   Glue the tags onto the poster board to represent the kinds and the relative abundances of the organisms observed.

Which colors represent producers?

medium green, red, light green, and yellow

Which colors represent consumers?

brown and blue

Which color represents decomposers?

orange

4. After the maps and observations have been completed, rewind the string and remove the stakes. Take nothing but photographs, and leave nothing but footprints.

5. Return to the classroom and place the site map on a bulletin board. Use the maps of the physical features and the biotic factors, and the physical features of the site to answer the Analysis questions below. The lab groups may need to supply information for the rest of the class about the specific kinds of plants and animals represented by the tags on the map.

Analysis

Use the information on the Knowing Side of your Vee Form to interpret your results and those of your classmates and then determine your **Knowledge Claim**. Write a **Value Claim** for this study. As you answer the following Analysis questions, you may wish to refer to the ecosystem pyramid that follows question 2.

1. List two food chains found on the maps that contain a producer, a primary consumer, and a secondary consumer.

   Answers will vary. (For example: grass, grasshopper, bird)

2. Which group, producers or consumers, contains the largest number of indiviuals?

Usually the producers will represent the largest number of individuals, but a small site with a large anthill and few plants may be an exception.

3. Which group, producers or the consumers, represents the largest group in biomass (volume and weight)?

Usually the producers will represent the largest group in biomass, but with a site that is small, animals could be in the study area, although they receive energy from another area.

4. What is the relationship between the quantity of biomass of the largest trophic level to the amount of energy available for transfer to higher levels?

At each trophic level, the available energy becomes less. The biomass tends to decrease along with the energy available for transfer to the next level.

5. Are the producers or the decomposers present in larger numbers?

Usually the producers will represent the group with the larger number on the map.
Bacteria, which are major decomposers, will not be visible even though they may be plentiful.

6. What role does the decomposer play in the environment?

The decomposer breaks down dead plants or animals and returns the nutrients to the soil.

7. What would happen to the producers and consumers if there were no decomposers?

They would die. The producers and consumers would soon be without vital nutrients if no decomposers were present.

8. Describe any difference between the observed biotic factors and the abiotic factors.

Answers will vary. For example, fewer plants may be found on sloping land where the soil

has eroded, or more plants may be found near a stream where water is plentiful.

_____

_____

_____

Going Further

Generate a **New Focus Question** that could be the point of a new investigation based on what you now know.

• Examine the influence of humans on the site studied in this investigation.

Mapping an Environmental Site

**Investigation 12-1
Vee Form Report Sheet**

DOING SIDE

Value Claim:

If you keep a map of an environmental site over the years and add to it every year, you can assess the effects of environmental changes.

Knowledge Claim:

Sample: The trees, bushes, animals, dead logs, and fungi are biotic factors, and the rocks, air, water, and soil are abiotic factors. The birds use the trees for nests. The insects lay egg cases in the grass, on the trees, and under rocks. The animals and fungi change the soil and the animals eat plants and each other.

Records:

The poster board on which the students record their observations serves as the Records section in this investigation.

Materials:

meter stick, stakes (4), string, poster board (square), large felt-tip pen, tags (240), glue

Procedure:

Measure a 100 square meter site and mark the perimeter with string and stakes. Draw the site and glue appropriate tags to the poster for keyed factors.

KNOWING SIDE

Focus Question:

What are the biotic and abiotic factors in our environmental site? How many of the organism types are there? What effects do biotic and abiotic factors have on each other?

New Focus Question:

How does our site compare to other sites nearby? What is the influence of human activity on the site studied?

Subject Area:

Ecosystems

Concepts:

ecology
food web
community
consumers
producers
biomass
decomposers

Vocabulary:

biotic
abiotic

Concept Statements:

1. Ecology is the study of the relationship of living things and their environment.
2. Organisms are the biotic factors.
3. The physical, non-living features are the abiotic factors.
4. The food webs of the community show how organisms are interrelated.
5. Decomposers break down dead organisms.
6. Producers make their food through photosynthesis and may be eaten by consumers who cannot synthesize their own food.
7. Producers show greatest numbers, biomass, and energy in ecological pyramids.

Additional Records and Observations:

Detergents as Pollutants

Introduction

Caution students to avoid dusting the air with detergent, in order to keep it out of their eyes and lungs.

The proper disposal of waste products has long been a problem for human populations. However, the human population has increased rapidly in recent years, making pollution control more urgent. Many cities treat contaminated water before releasing it into rivers, lakes, or oceans, but some remaining chemical substances are still released with the treated water. Phosphates are a major source of pollution and are most often released in this manner. Detergents are the most common source of phosphates. The nutrient enrichment of the water caused by adding chemicals, including phosphates, is called *eutrophication*. In this investigation, you will determine the effects of detergents on an aquatic ecosystem.

Materials

- Aquaria (3)
- Pond water
- Wax pencil
- Detergent with phosphate, 2 g
- Detergent without phosphate, 2 g
- *Spirogyra*, stock culture
- Paper towels
- Balance
- Plastic wrap
- Fluorescent lamp

Focus Question

Design your own **Focus Question** based on the information presented in the Introduction and Materials sections above, as well as the additional information presented in the Procedure section that follows. Write your Focus Question on the Vee Form.

Knowing Side

With the help of your teacher, your class will discuss the **Subject Area** background for this investigation. Then, with the help of your classmates, list the **Concepts** and the new **Vocabulary** words on your Vee Form. In the **Concept Statements** section of your Vee Form, use these words in sentences that define and explain them.

Procedure

Spirogyra is suggested for this investigation because it is readily found and easily handled. Other types of algae will also give successful results and may be found in nature or obtained from a biological supply company.

Pond water is preferable for use in this investigation, but if it is not available, tap water that has been aged 24 hours can be substituted.

1. Fill each aquarium two-thirds full of pond water. Label the aquaria #1, #2, and #3. To aquarium #1, add 2 g of detergent with phosphates. To aquarium #2, add 2 g of detergent without phosphates. To aquarium #3, add no detergent.

What is the purpose of aquarium #3?

It acts as a control.

_____

_____

2. Remove the *Spirogyra* from the culture container and place it briefly on a folded paper towel to absorb the excess water. Using the balance, weigh out three identical samples of *Spirogyra*. Record the initial mass of each sample in Table II on the back of your Vee Form. Add one *Spirogyra* sample to each aquarium. Be sure the samples are large enough to cover much of the aquarium, as shown in Figure I on the following page. This procedure must be done quickly to keep the *Spirogyra* from drying out.

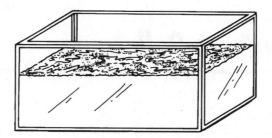

Figure 1

What is the reason for weighing each sample that is placed in the aquaria?

Identical samples make the presense of detergent the only variable.

3. Cover each aquarium with a sheet of plastic wrap and place all of them 20 cm from a fluorescent lamp.

What is the purpose of the lamp?

to provide light energy for photosynthesis

4. Observe the aquaria twice each week, and record your observations in Table I in the **Additional Records and Observations** section on the back of the Vee Form. Any other details you feel should be noted should be included in the **Records** section on the front of the Vee Form.

5. After three weeks, remove the *Spirogyra* from each aquarium and briefly place each sample on a folded paper towel to absorb the excess water. Weigh the *Spirogyra* found in each aquarium and record its mass in Table II on the back of the Vee Form.

6. Find the *Spirogyra* mass increase for each aquarium and record these in Table II.

7. Clean up your materials and wash your hands before leaving the lab.

Use the information on the **Knowing Side** of your Vee Form to interpret your results from the **Doing Side**, and then determine your **Knowledge Claim**. Write a **Value Claim** for this study.

Analysis

1. Which aquarium had the greatest increase in *Spirogyra* mass? Explain.

The greatest increase in mass will probably be found in the aquarium with the detergent

containing phosphate. Algae thrive with this nutrient present. However, other detergents may provide nutrients that will also promote growth.

2. How did the color vary in each of the aquaria?

The algae in the aquaria containing detergent should appear greener. However, as the

algae grow and start to become overcrowded, the available light may be blocked by the

Spirogyra, and the algae may appear unhealthy and brownish, particularly near the

bottom.

3. In what part of the aquarium was the *Spirogyra* found?

It floated up to the top.

4. Why do you think the *Spirogyra* was found in that part of the aquarium?

As photosynthesis takes place, oxygen bubbles are produced. They are trapped in the

strands of *Spirogyra*, causing the *Spirogyra* to float.

5. If the *Spirogyra* shows very rapid growth in a lake, pond, or river, what effect do you think it would have on the other organisms in the aquatic system?

Since the *Spirogyra* floats, light is blocked and may not reach other organisms. Later, the

dying algae cause oxygen to be depleted as they decay. Other organisms die, since they

cannot survive without an adequate oxygen supply.

6. Which aquarium best displays this rapid type of growth?

the aquarium with the phosphate detergent, most probably

7. How can you account for the rapid growth?

Phosphates are an excellent nutrient for algae.

8. What are the characteristics of eutrophication?

The characteristics of eutrophication are increased growth of algae and a dark green
color. As the algae die and decay, there is a decrease in available oxygen, which can
cause the death of other aquatic organisms such as fish.

_____

_____

_____

_____

9. If you were a conservationist, what actions would you suggest to maintain aquatic ecosystems?

Students might suggest passing laws to prevent or control pollution and contamination,
using better methods of purifying polluted water, or advocate using detergents that do
not add nutrients that cause eutrophication to the water.

_____

_____

Going Further

Generate a **New Focus Question** that could be the point of a new investigation based on what you now know.

Name _____ Date _____ Class _____

Detergents as Pollutants

DOING SIDE

Value Claim:

Knowledge of the effects of phosphates on aquatic systems should stimulate the creation of anti-pollutant laws regarding detergents with phosphates.

Knowledge Claim:

The greatest increase in mass of algae is found in the phosphate aquarium. Initially, it was very green, but the bottom parts received no light and turned brown. The bubbles of oxygen from photosynthesis eventually disappeared as decay set in.

Records:

[See Additional Records and Observations on the following page]

Materials:

pond water, 1-gal. aquaria (3), wax pencil, detergent with phosphate (2g), detergent without phosphate (2g), Spirogyra, paper towels, balance, plastic wrap, fluorescent lamp

Procedure:

Create an experimental system to investigate the effects of detergent as a pollutant by setting up aquaria with phosphate-containing detergent, phosphate-free detergent, and no detergent.

KNOWING SIDE

Focus Question:

What is the effect of detergent as a pollutant on algae in an aquatic ecosystem?

New Focus Question:

Are all detergents alike, so that we can make generalizations about them? Do different detergents have different effects on aquatic systems?

Subject Area:

Ecosystems

Concepts:

pollution

Vocabulary:

eutrophication
phosphates

Concept Statements:

1. Waste disposal problems are urgent.
2. Phosphates are a major source of pollution and are released with treated water.
3. Detergents are the most common source of phosphates.
4. Eutrophication is the nutrient enrichment of the water caused by adding chemicals such as phosphates.
5. Phosphates are not in all detergents.

Additional Records and Observations:

Table I

Obs.	Aquarium #1 Color of Spirogyra	Odor of Aquarium	Presence of Bubbles	Position of Spirogyra	Obs.	Aquarium #2 Color of Spirogyra	Odor of Aquarium	Presence of Bubbles	Position of Spirogyra	Obs.	Aquarium #3 Color of Spirogyra	Odor of Aquarium	Presence of Bubbles	Position of Spirogyra
1					1					1				
2					2					2				
3					3					3				
4					4					4				
5					5					5				

Table II

Mass of Spirogyra

After Week Three: _____ g

Initial Mass: _____ g

Mass Increase: _____ g

Mass of Spirogyra

After Week Three: _____ g

Initial Mass: _____ g

Mass Increase: _____ g

Mass of Spirogyra

After Week Three: _____ g

Initial Mass: _____ g

Mass Increase: _____ g

Acid Rain Effects

Introduction

Acid rain is one of the most harmful forms of pollution; it is caused by the burning of fossil fuels. It forms when sulfur and nitrogen oxides dissolve in water vapor in our atmosphere. These dissolved oxides form sulfuric and nitric acids, which fall to the ground contained in precipitation. A major source of acid rain is the burning of high-sulfur coal.

All precipitation is slightly acidic, but when it is highly acidic it can damage living organisms, buildings, bridges, etc. The low pH of the rain also promotes leaching of minerals from the soil, and kills fish and other aquatic organisms. Because acid-laden moisture collects in clouds and moves with the wind, acid sources in one place may cause acid pollution from rain or snow in another place hundreds of miles away.

In this investigation you will demonstrate the effect of acid rain on metal and fruit and observe the effect of acid rain on seeds.

Materials

To make synthetic acid rain, add 5.5 mL of 0.01M H_2SO_4 to 1 L of water. The pH of the resulting solution is close to 4.

- Acid rain
- Wax pencil
- Small beakers (4)
- pH paper test strips
- Petri dishes (2)
- Cotton
- Filter paper to fit petri dishes

- Radish or mustard seeds
- Steel wool
- Red apple peel
- Graduated cylinder

Focus Question

Design your own **Focus Question** based on the information presented in the Introduction and Materials sections above, as well as the additional information presented in the Procedure section that follows. Write your Focus Question on the Vee Form.

Knowing Side

With the help of your teacher, your class will discuss the **Subject Area** background for this investigation. Then, with the help of your classmates, list the **Concepts** and the new **Vocabulary** words on your Vee Form. In the **Concept Statements** section of your Vee Form, use these words in sentences that define and explain them.

Procedure

The set-up day takes about 25 minutes. The observation day takes about 30 minutes if analyses are completed in class.

1. **CAUTION: The synthetic acid rain is a skin irritant. Put on goggles, gloves, and a lab apron during steps 2–8.**
2. Observe your teacher make synthetic acid rain in class.
3. Label one beaker as "water" (use tap water) and one as "acid rain." Put 25 mL of each liquid into the appropriate beaker.
4. Test the liquid in each other with a pH paper strip and record the pH of each liquid in the **Records** table of your Vee Form.
5. In each of the beakers, place a small piece of red apple peel and a bit of steel wool, and leave overnight.
6. Label the third and fourth beakers as water and acid rain, and put 25mL of the appropriate fluid in both, but put 50 seeds in each and leave overnight.
7. After 24 hours, note any changes in the items soaked overnight.

8. Prepare germination petri dishes by spreading a layer of cotton in the bottom of two dishes and covering the cotton with a circle of filter paper. Add the liquid and seeds from the third and fourth beakers to two different dishes and label the contents. Cover the petri dishes and leave overnight.
9. Clean up your materials and wash your hands before leaving the lab.
10. After 24 hours, count the number of seeds that have germinated in each dish and record these numbers in the table on the Vee Form. Express this result as a percentage of the total number of seeds planted, then write this value in the Vee table.

All leftover acid rain should be poured into a common container and neutralized to pH 7 by adding a dilute base dropwise before pouring it down the drain.

Use the information on the **Knowing Side** of your Vee Form to interpret your results from the **Doing Side**, and then determine your **Knowledge Claim**. Write a **Value Claim** for this study.

Analysis

1. Why is it relevant to look at the effects of acid rain on steel wool and on an apple peel in an investigation about acid rain and seed germination?

   Acid rain affects bridges, buildings, and other structures as well as living things.

2. What do you predict the pH of normal rain to be?

   Normal rainwater is slightly acidic (pH about 5.6), due to the formation of carbonic acid in rainwater when CO_2 combines with water.

3. Why is the effect of acid rain on steel important to human beings?

   Steel is corroded by acid rain and hence weakened.

4. What was the effect of acid rain on the seeds? Of what importance is this result?

   Acid rain seems to hinder germination and is thus damaging to plants as well as animals.

Going Further

Generate a **New Focus Question** that could be the point of a new investigation based on what you now know.

Name _____ Date _____ Class _____

Acid Rain Effects

KNOWING SIDE

DOING SIDE

Focus Question:

What is the effect of acid rain on steel and living plants?

Value Claim:

Acid rain could harm the agricultural industry, and damage bridges, buildings, and other structures.

Knowledge Claim:

Acid rain may prevent the germination of seeds or slow down plant growth because of cell damage.

New Focus Question:

Are there some seeds that would germinate better under slightly acidic conditions?

Records:

Seeds Germinated		
Medium	Number	%
Normal H₂O pH (6–7)		
Acid Rain pH 4		

Subject Area:

Theory of environmental cycles
Pollution theory

Concepts:

acid rain
pH
fossil fuel burning

Vocabulary:

germination

Materials:

Acid rain, wax pencil, small beakers (4), pH paper test strips, petri dishes (2), cotton, filter paper, radish or mustard seeds, steel wool, graduated cylinder.

Procedure:

Compare seed germination rates in both water and acid rainwater. Examine the effects of acid rain on both metal and an apple peel.

Concept Statements:

1. Acid rain results from the burning of fossil fuels as the nitrogen and sulfur oxides combine with water vapor in the air, forming nitric acid and sulfuric acids. This falls to the earth contained in rain.

2. Acid rain lowers the pH of rivers and lakes and affects aquatic organisms.

3. Seed germination is characterized by the embryonic plant breaking out of the seed coat.

4. pH is a measure of acidity. The lower the pH, the more acidic a solution is.

Additional Records and Observations:

Oil-Degrading Microbes

Introduction

Increasingly, the public has been made aware of oil spill hazards through the media. A single gallon of oil can spread thinly enough to cover up to four acres of water. Some evaporates and some emulsifies, to form a heavy material that sinks to the bottom of the ocean. This is what endangers birds and marine mammals. Oil that doesn't emulsify is broken down by microorganisms and radiant energy.

The countermeasures for oil spills, other than prevention, include the mechanical use of skimmers and barriers, chemical dispersants and solvents, and the burning of oil. However, there are microorganisms that can break down the different hydrocarbons in the oil spill; they may be the best, most environmentally safe prospect for the clean-up of oil spills. These "oil hungry" bioremediation microbes convert oil into food and non-toxic living cells, which then become assimilated safely in the aquatic food chain.

In this investigation you will compare the physical characteristics of oil before and after action by oil-degrading microbes. Also, you will compare the growth and oil-degrading ability of a single bacterial with that of a single fungal strain.

Although the oil-degrading microbes are naturally occurring and nonpathogenic, it is recommended that you use goggles and gloves throughout the procedures to minimize the risk of any contamination.

Materials

- Test tubes with caps (2)
- Distilled water, 60 mL
- Density indicator strip (Ward's)
- Test tube rack
- *Pseudomonas* culture
- *Penicillium* culture
- Refined oil

Focus Question

Design your own **Focus Question** based on the information presented in the Introduction and Materials sections above, as well as the additional information presented in the Procedure section that follows. Write your Focus Question on the Vee Form.

Knowing Side

With the help of your teacher, your class will discuss the **Subject Area** background for this investigation. Then, with the help of your classmates, list the **Concepts** and the new **Vocabulary** words on your Vee Form. In the **Concept Statements** section of your Vee Form, use these words in sentences that define and explain them.

Procedure

You need to rehydrate and grow both cultures four to five days prior to use in the investigations. You may also have your students participate in the procedures of rehydration and growing of cultures. Rehydration instructions come with the order of freeze-dried

1. **CAUTION: Put on a lab apron, safty goggles, and disposable gloves.**
2. Label your test tubes for each species of microorganism. Add 5 mL of distilled water to each tube.
3. Add 4–5 drops of refined oil, so as to form a thin layer in each test tube. Since the tubes do not contain any other nutrients, the microbes will use the oil as a source of carbon.
4. Inoculate the appropriate test tubes with 5 mL of different cultures, one of the bacterial strain and one of the fungal strain.
5. Attach the cap on each tube and invert it several times to mix the oil with the microorganisms. This wave action would occur in the ocean and increases the dissolved oxygen concentration in the water.

Pseudomonas and *Penicillium* cultures from Ward's or other scientific supply houses.

6. Incubate the test tubes in racks with caps half loosened at 30°C, or place them in a warm spot in your lab room if you have no incubator.
7. Look for any signs of oil degradation, such as the formation of tiny oil droplets, breakup of the oil layer into smaller fragments, or changes in texture.
8. Note the color of the oil.
9. Place the density indicator strip vertically, below the water level, with the darkest bar at the top of the panel on one of the test tubes. Number the bars from 1 to 5, with 5 for the darkest. Measure the turbidity by recording the number of bars on the panel that can no longer be clearly seen when the tube is held up to a white background. Remove the density indicator strip, washing it before placing it in the other tube, and repeat the procedure.
10. Invert the tubes once or twice daily to increase the dissolved oxygen content in the water and further mix the microbes with the oil. Observe your tubes once every 24 hours for 3–4 days. Growth will mainly be observed at the interface between oil and water.

In the **Additional Records and Observations** section table on the back of your Vee Form, record your observations of general appearance characteristics of the degrading oil, its color, and the turbidity of the water.

Analysis

1. Describe the initial physical characteristics and appearance of oil for each tube.

    The refined oil used in this investigation is light brown in color and forms a smooth and
    continuous layer on the water surface.

    _____

    _____

    _____

    _____

2. Describe any changes in the physical characteristics and appearance of the oil on Day 1 and beyond and discuss the possible causes for such changes.

The oil starts to show signs of degradation after 1 or 2 days of microbial action, based on its color change from a light brown to an ivory color. Another change in its physical characteristics is the breakdown of the continuous layer of oil on the water surface into minute oil droplets.

3. Is there a difference in the rate of oil degradation between the bacterial and fungal cultures?

There does not appear to be a difference in degradation rates between the bacterial and fungal cultures. However, over time there is a difference in the growth pattern. Both microorganisms start to grow at the interface between oil and water. However, the fungus grows in mats (mycelial mass), which will form over the oil surface.

4. Can you think of any advantages of using the bacteria over the fungus to degrade oil?

Each microorganism tends to be specific in the types of hydrocarbons it will degrade. Thus, there is no real advantage in using a bacteria over a fungus to degrade oil, other than in specific situations. However, by using a mixture of microorganisms, a broad spectrum of hydrocarbon degradability that is not possible when using only one type of microorganism can be attained.

5. What does an increase in turbidity indicate?

An increase in turbidity indicates an increase in microbial population as well as degradation of the oil.

6. What is the turbidity level of your cultures after four days of incubation? How long do you think your cultures will continue to grow?

The bars (2–3) on the density indicator strip should disappear after 4 days of incubation. The cultures should continue to grow until oxygen and the carbon source are depleted.

You may wish to purchase density indicator strips from Ward's.
(15 M 1996) Pkg/25
(15 M 1998) Pkg/10

7. What is the limiting growth factor in your test tube?

The presence of oil is the major limiting growth factor, since it is the only carbon source
added in the test tubes. The dissolved oxygen level is another limiting factor, since these
microorganisms are aerobic.

_____

8. Clean up your materials and wash your hands before leaving the lab.

Going Further

Generate a **New Focus Question** that could be the point of a new investigation based on what you now know.

Each team of 2 to 4 students should test the oil-degrading ability of the cultures.

Set up a culture-growing station with a plastic pipette in each one of them, for students to inoculate their test tubes with.

Although specific observations may vary from student to student, they all should agree that the oil started showing signs of degradation after 1 to 2 days of microbial action, based on its color change from a light brown to an ivory color and changes in the physical characteristics. The turbidity of the water should increase, indicating microbial growth. However, oil is the limiting growth factor, since it is the only carbon source added in the test tube.

Name _____ Date _____ Class _____

Oil-Degrading Microbes

KNOWING SIDE

DOING SIDE

Subject Area:
Environmental cleanup

Focus Question:
What are the physical characteristics of oil before and after action by the oil-degrading microbes? Which is a better biodegrader of oil, a bacterial strain or a fungal strain?

Concepts:
biodegradation

Vocabulary:
emulsification
turbidity

New Focus Question:
Do various oil-degrading microbes work more or less effectively, depending on the grade of crude oil?

Value Claim:
By using a mixture or microorganisms, a broader spectrum of degradability of hydrocarbons is attained, so more than one should be probably be used in the mixture.

Knowledge Claim:
Both microorganisms degrade oil at the same rate. Fungus grows over the oil surface where it appears white and fuzzy. Bacterial growth is yellow and is also raised above the surface. Each organism might be specific to the type of hydrocarbon it degrades.

Records:
[See Additional Records and Observations on the following page]

Materials:
Test tubes with caps (2), 60 mL distilled water, density indicator strip, test tube rack, *Pseudomonas* culture, *Penicillium* culture, refined oil.

Procedure:
Test the oil-degrading ability of two naturally occurring strains of *Penicillium* and *Pseudomonas*.

Concept Statements:
1. Oil spillage from various sources is an increasing problem.
2. A gallon of oil spreads over four acres of water, losing 25% to evaporation and 60% to emulsification.
3. Microorganisms provide an alternative to conventional means to combat oil spills.
4. There are short- and long-term effects of oil spills.

Additional Records and Observations:

Day	Species	General Appearance of Oil	Color of Oil	Turbidity of Water
0	Pseudomonas			
0	Penicillium			
1	Pseudomonas			
1	Penicillium			
2	Pseudomonas			
2	Penicillium			
3	Pseudomonas			
3	Penicillium			
4	Pseudomonas			
4	Penicillium			

Classification

Introduction

The five-kingdom system of classification of organisms was proposed in 1959, and they are the five categories most biologists use today. Viruses are not classified as living things; they lack a cell structure, and so are not included in any kingdom. Scientific names are Latin and are very specific to their species. This allows scientists to refer to a particular organism in such a way that other scientists will always recognize it as being the same organism.

In Part A of this investigation, you will classify several items by observing them and looking for similar features by which to group them. In Part B, you will attempt to classify some familiar organisms. You will observe only drawings of your organisms, so your classification in this investigation will depend upon physically observable traits alone.

Materials

20 items such as: leaves, acorns, various nut types, twigs, rocks, pebbles, pine needle bundles, pine cones, rocks, pebbles, paper clips, nails, screws, soil, chalk, coins, toothpicks, thumbtacks, etc.

Focus Question

Design your own **Focus Question** based on the information presented in the Introduction and Materials sections above, as well as the additional information presented in the Procedure section that follows. Write your Focus Question on the Vee Form.

Knowing Side

With the help of your teacher, your class will discuss the **Subject Area** background for this investigation. Then, with the help of your classmates, list the **Concepts** and the new **Vocabulary** words on your Vee Form. In the **Concept Statements** section of your Vee Form, use these words in sentences that define and explain them.

Procedure

Part A: Classification of Familiar Objects

1. Examine all of your items closely and try to determine their composition. Also note how they all differ or are similar in size, shape, weight, color, or any other features you may notice. Devise a classification scheme that will allow you to divide the items into two groups, Kingdom I and Kingdom II. List and describe the characteristics of the items in your two kingdoms in the appropriate spaces in Table 1 of the **Additional Records and Observations** section on the back of your Vee Form.
2. Now study the objects again and find another characteristic that can divide Kingdom I into two phyla. Write a descriptive title under the appropriate phylum section (Phylum IA or IB), list the items that now are included in each phylum, and count up the total number of phylum objects. Do the same for Phylum IIA and IIB.
3. Now examine each of your four subgroups and find some characteristic that enables you to divide them up into your final eight groups on the data table. Again create descriptive labels for Class IA$_1$, Class IA$_2$, Class IB$_1$, Class IB$_2$, Class IIA$_1$, Class IIA$_2$, Class IIB$_1$, and Class IIB$_2$.

4. Add the total number of items in each of your subgroupings on the proper lines in the data table.

Part B: Classification of Familiar Animals

5. As you look at Figure 1, determine how these animals are similar to, and different from, each other. In Table 2 on the back of your Vee Form, split up the animals into classified groupings as you did for the items in Part A, giving descriptive names for each category.

Figure 1

Use the information on the **Knowing Side** of your Vee Form to interpret your results from the **Doing Side**, and then write your **Knowledge Claim.** Write a **Value Claim** for this study.

Analysis

1. What features of the items in Part A were helpful in grouping them?

Answers will vary.

_____

_____

_____

_____

2. Check your text to find the true kingdom, phylum, and class names for the animals in Part B.

All are in Kingdom Animalia; the snail (Class Gastropoda) and clam (Class Bivalvia) are

both in Phylum Mollusca; the spider and tick (Class Arachnida), the roach and bee (Class Insecta) and the lobster (Class Crustacea) are all in Phylum Arthropoda; the frog and

salamander (Class Amphibia), the mouse, horse, and human (Class Mammalia), and the bird (Class Aves) are all in Phylum Chordata, Subphylum Vertebrata.

_____

_____

_____

_____

Going Further

In the appropriate place on your Vee Form, write a **New Focus Question** that could be the point of a new investigation based on what you now know.

Name _____ Date _____ Class _____

Classification

KNOWING SIDE

DOING SIDE

Focus Question:

How can many different nonliving and living objects be classified?

Subject Area:

Taxonomy
Nomenclature

New Focus Question:

If a new organism was discovered and it didn't fit any existing categories, what would scientists do?

Concepts:

classification

Vocabulary:

taxonomy
kingdom
phylum
class

Value Claim:

Classification of organisms is useful to scientists as they attempt to develop new medicines to combat disease.

Knowledge Claim:

Living things may be classified by the characteristics they display into groups and subgroups. The more specific the group characteristics are, the fewer organisms are in the group.

Records:

[See Additional Records and Observations on the following page]

Materials:

Answers will vary according to items selected.

Procedure:

Group items in general categories and then break down those categories into more and more specific ones.

Concept Statements:

1. A classification of items enables one to group them by common characteristics.
2. Classification of species displays evolutionary relationships.
3. By observing and examining organisms, scientists have developed a classification scheme for them.
4. Organisms are placed in taxonomic groups according to similar features or traits.

Additional Records and Observations:

Table 1

All Items: _____

Kingdom I: _____			Kingdom II: _____		
No. of items: _____			**No. of items:** _____		
Phylum IA: _____ No. of items: _____		Phylum IB: _____ No. of items: _____	Phylum IIA: _____ No. of items: _____		Phylum IIB: _____ No. of items: _____
Class IA$_1$: No. of items: _____	Class IA$_2$: No. of items: _____	Class IB$_1$: No. of items: _____ Class IB$_2$: No. of items: _____	Class IIA$_1$: No. of items: _____	Class IIA$_2$: No. of items: _____	Class IIB$_1$: No. of items: _____ Class IIB$_2$: No. of items: _____

Table 2

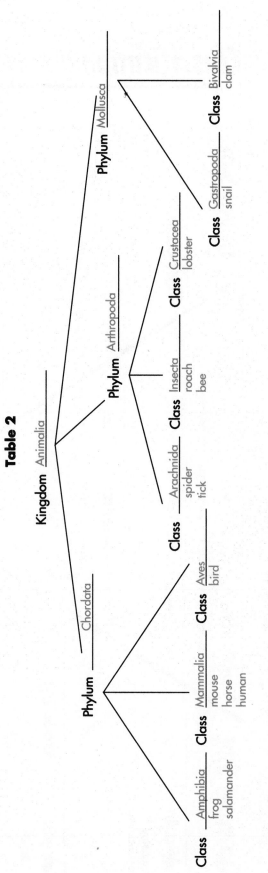

Antibiotics and Zones of Inhibition

Introduction

Review the microbiological safety section in the teacher's manual before conducting this lab.

Commercial dehydrated broth is available and comes with directions for preparation on the bottle. See a biological supply catalog.

In 1929, Sir Alexander Fleming observed that a mold contaminant prevented or inhibited the growth of microorganisms in a petri plate. Through following up on this observation, he greatly advanced medical science. Many antibiotics are the byproducts or secretions of molds that are commonly found in soil, air, and water.

Penicillium is a blue-green mold that is often found on decaying fruit. The antibiotic ingredient in *Penicillium* prevents some bacteria from properly building their cell walls. As a result, these bacteria usually burst from internal pressure or are readily destroyed by the body's immune system. In this investigation, you will determine the effect of *Penicillium* and other antibiotics on various types of bacteria.

Clear areas on the culture plate near the antibiotic disks you will place there indicate a lack of bacterial growth. These areas are called zones of inhibition.

Materials

- Wax pencil
- Petri plates
- *E. coli*, stock culture
- *B. subtilis*, stock culture
- Sterile cotton swabs
- Antibiotic sterile disks
- Control paper disks
- Incubator
- Forceps
- Metric ruler

Focus Question

Design your own **Focus Question** based on the information presented in the Introduction and Materials sections above, as well as the additional information presented in the Procedure section that follows. Write your Focus Question on the Vee Form.

Knowing Side

With the help of your teacher, your class will discuss the **Subject Area** background for this investigation. Then, with the help of your classmates, list the **Concepts** and the new **Vocabulary** words on your Vee Form. In the **Concept Statements** section of your Vee Form, use these words in sentences that define and explain them.

Procedure

Prepare the plates with nutrient agar prior to the investigation. Prepoured plates are available from biological supply companies.

Emphasize the technique for keeping the plates sterile to your students. It is very critical to the success of the lab investigation results.

When the students streak the agar surface, remind them to streak the <u>entire</u> surface using an s-motion for better results.

1. **CAUTION: Put on a lab apron, goggles, and disposable gloves.**

2. Obtain two petri plates containing sterile nutrient agar. Label one *E. coli* and the other *B. subtilis*. Also label the plates with your initials.

3. Streak one bacterial culture, covering its petri plate in a continuous s-shaped motion, using a cotton swab. When you remove the cap from the broth tube, do not put the cap on the table, but continue to hold it between your fingers, so as not to contaminate the tube contents or anything in the lab. Do the same for your second petri plate, using the other bacterial culture.

If no incubator is available, room temperature for 48–72 hours will suffice.

Use proper caution when disposing of bacteria.

Placing the plates upside down prevents condensation from dripping onto the culture.

4. From your teacher, obtain one control paper disk, along with one of each of three antibiotic disks. Each disk will be placed in a quadrant of one of your plates, using the forceps. Number the quadrants on the outside of the bottom of the petri plate and carefully place the disks onto the center portion of each quadrant. Be sure to record which antibiotics were placed in each of the numbered quadrants in Table 1, which is provided in the **Additional Records and Observations** section on the back of your Vee Form.

5. Repeat steps 3 and 4 for the other bacterial petri plate. Record your antibiotic types in Table 2 on the back of your Vee Form as before.

6. Tape the plates shut and place them upside down in the incubator at 37°C overnight.

7. After incubation, check the growth in each plate. Measure the size (diameter) in millimeters of any zone of inhibition with your metric ruler and record your results in Tables 1 and 2. Do not open plates.

8. Continue to incubate and check the cultures after 48 and 72 hours, if time permits.

9. Clean up your materials and wash your hands before leaving the lab.

Use the information on the **Knowing Side** of your Vee to interpret your results from the **Doing Side**, and then write your **Knowledge Claim**. Write a **Value Claim** for this study.

Analysis

1. How sensitive is each bacterium to each of the various antibiotics? How was this determined?

   Answers will vary. This was determined by measuring the size of zones of inhibition around the various disks.

2. What experimental proof do you have that the paper disk itself cannot stop bacterial reproduction?

   There was absolutely no zone of inhibition around the control disk, for it contained no antibiotic. The bacteria also grew under the disk.

3. If you were sick with a sore throat and the doctor suspected a bacterial infection, he or she probably wouldn't wait 24 hours while a throat culture was taken and incubated. Instead, a broad spectrum antibiotic (a mixture of antibiotics) would be prescribed immediately and the culture results evaluated the next day to be sure the medication was correct. If, after 24 hours the doctor changed your antibiotic medication, what could be the reason?

   The doctor would know what the specific bacteria is and could prescribe an antibiotic that is most effective in fighting the infection.

Going Further

In the appropriate place on your Vee Form, write a **New Focus Question** that could be the point of a new investigation based on what you now know.

Antibiotics and Zones of Inhibition

KNOWING SIDE

DOING SIDE

Focus Question:

What is the effect of different antibiotics on various bacterial cultures?

Subject Area:

Disease

New Focus Question:

Which antibiotic mixture has the broadest spectrum effect on various bacteria?

Concepts:

zone of inhibition
antibiotic
incubation

Vocabulary:

bacterial culture

Value Claim:

Not all antibiotics are equally effective against different infections. Those that most readily combat the specific infection should be prescribed.

Knowledge Claim:

The disks showed the antibiotics with the greatest inhibition zones to be the most effective on those particular bacterial cultures.

Records:

[See Additional Records and Observations on the following page]

Concept Statements:

1. The zone of inhibition is an area lacking growth by a bacterial strain/culture, caused by the presence of an antibiotic.
2. Antibiotic substances can prevent bacteria from reproducing.
3. Many bacteria grow best in a warm environment that an incubator can provide.

Materials:

wax pencil, petri plates (2), *E. coli* stock culture, *B. subtilis* stock culture, cotton swabs, antibiotic sterile disks, control paper disks, incubator, forceps, metric ruler.

Procedure:

Place 3 antibiotic disks and a control disk on the 4 different quadrants of 2 petri plates inoculated with different bacteria. Grow in an incubator overnight and measure zones of inhibition.

Additional Records and Observations:

Table 1. Antibiotic Effects on *E. coli*

Petri Dish Quadrant	Antibiotic Disk	Zone of Inhibition Diameter (mm)
1		
2		
3		
4	Control	None

Table 2. Antibiotic Effects on *B. subtilis*

Petri Dish Quadrant	Antibiotic Disk	Zone of Inhibition Diameter (mm)
1		
2		
3		
4	Control	None

Bactericidal Effect of Soap

Introduction

One of the best ways of preventing the spread of bacterial and viral infections is to frequently wash your hands with soap and water. Many soaps advertise that their ingredients will destroy bacteria normally found on the body. In this investigation, you will discover how effectively soaps can kill bacteria, and examine the relationship between handwashing techniques and bactericidal effectiveness.

Materials

- Wax pencil
- New scrub brush
- Petri plates with sterile nutrient agar (5)
- Liquid antibacterial soap
- Incubator
- Cellophane tape

Focus Question

Design your own **Focus Question** based on the information presented in the Introduction and Materials sections above, as well as the additional information presented in the Procedure section that follows. Write your Focus Question on the Vee Form.

Knowing Side

With the help of your teacher, your class will discuss the **Subject Area** background for this investigation. Then, with the help of your classmates, list the **Concepts** and the new **Vocabulary** words on your Vee Form. In the **Concept Statements** section of your Vee Form, use these words in sentences that define and explain them.

Procedure

Be sure to choose a soap that is an antibacterial soap. If you would like to add another facet to this investigation, add a liquid soap not touted to be antibacterial. Then you must add another petri plate per group.

Day 1

1. While keeping the petri plates closed at all times, use the wax pencil to label the bottoms of four petri plates with: "control," "no soap," "1 minute,"and "4 minutes."
2. Without washing your hands, carefully press different surfaces of your hand on the agar in the petri plate marked "control." Remember, you should treat all bacterial culture media as if they contain pathogens.
3. Use the scrub brush to scrub all surfaces of your hands under running water for 1 minute without using any soap. Spend the same amount of time on each hand (30 seconds) and scrub under the fingernails. When you are finished, do not dry your hands. Have a partner turn off the water faucet and remove the lid of the petri plate for you. Without touching anything first, carefully press different surfaces of your hand on the plate marked "no soap," as you did for the control.
4. Repeat step 3 of this investigation using liquid soap as you scrub your hands for 1 minute (30 seconds for each hand). Make sure you do not touch anything before pressing different parts of your hand on the petri plate.
5. Repeat step 3 again, using liquid soap for 4 minutes of handwashing. Be sure to scrub each hand for 2 minutes.
6. Fasten the lid of each petri plate to its bottom half with two small pieces of cellophane tape. Place the plates upside down. Incubate all four plates over-night at 37°C.

Day 2

7. Remove the plates from the incubator and turn them right side up. Without removing the lid, check each plate for the presence of bacterial colonies, and count the number of colonies present in each plate. Return the plates, closed, to your teacher.
8. In the table in the **Additional Records and Observations** section of your Vee Form, record the growth of colonies present on each of the plates.
9. Clean up your materials and wash your hands before leaving the lab.

Use the information on the **Knowing Side** of the Vee to interpret your results from the **Doing Side**, and then write your **Knowledge Claim**. Write a **Value Claim** for this study.

Analysis

1. Compare the bacterial growth on the plates. Which plate had the most growth? Which had the least?

the control plate; the 4-minute plate

2. Does water alone effectively kill bacteria? Explain.

No; it does not have any bactericidal effects. Students will probably observe bacterial colonies on the plates marked "no soap."

3. How long must you scrub your hands with soap to kill bacteria?

Students will probably observe that soap does not have any significant bactericidal effects until their hands are washed for 4 minutes.

4. Compare your results with those of your classmates. What could account for the differences in the bactericidal effects of the soap?

Different people have different kinds of bacteria on their skin. Some students may have inadvertently touched something after washing their hands, but before they inoculated their plates.

Before this investigation, prepare five petri plates with nutrient agar for each student or group of students. Sterile nutrient agar may be purchased from Ward's.
Nutrient Agar (bottle) (88 M 1500)

Going Further

Generate a **New Focus Question** that could be the point of a new investigation based on what you now know.

Disposable plates are suggested for this investigation. If they are not available, autoclave or steam-sterilize the glass petri plates at 121°C at 15 psi for 15-20 minutes when the lab is over, and discard any used agar in a sealed metal or plastic container. Alternatively, flood or immerse the plates in chlorine laundry bleach for 30 minutes and then discard the used agar-bleach solution in a sealed container. There is no way to predict the type of bacteria that may be present on a person's hands, so care should be taken in the disposal of plates.

Bactericidal Effect of Soap

KNOWING SIDE

DOING SIDE

Focus Question:

How effectively do soaps advertised as bactericidal actually kill bacteria? Do various washing techniques differ in their bactericidal effects?

New Focus Question:

How would different brands of soap compare in bactericidal effects, especially to those that claim to kill bacteria?

Subject Area:

Disease theory

Concepts:

bactericide
control
antibacterial soap

Vocabulary:

colonies

Concept Statements:

1. Soaps are often advertised as being bactericidal.
2. One of the best ways to prevent the spread of bacterial and viral infections is to wash your hands often with soap and water.
3. The control is the setup to which all other plates can be compared.
4. Colonies are growing populations of bacteria.

Value Claim:

When a health professional visits a patient, s/he should always scrub with an antibacterial soap for about 4 minutes.

Knowledge Claim:

Bactericidal soaps are more effective if you wash with them for 4 minutes rather than only 1 minute.

Records:

[See Additional Records and Observations on the following page]

Materials:

wax pencil, new scrub brush, sterile petri plates (4) with nutrient agar, antibacterial liquid soap, incubator

Procedure:

Set up four different situations in petri plates with varied handwashing techniques using only water, water and soap, scrubbing for 1 minute and 4 minutes.

Additional Records and Observations:

Bactericidal Effect of Soap

Plate	Type of Soap	Duration of Handwashing	Growth of Colonies/ General Observations
1	none	none	most growth
2	water w/o soap	1 minute	growth
3	liquid soap	1 minute	growth
4	liquid soap	4 minutes	least growth

Observing Pond Water

Introduction

A pond or lake is an excellent source for a large variety of microscopic life forms. Innumerable protozoa, as well as many types of invertebrates, are easily obtained in water samples taken near the shoreline. In this investigation you will observe some of these freshwater microorganisms.

Materials

- Pond water
- Culture dish
- Wax pencil
- Slides (3)
- Medicine dropper
- Coverslips (3)
- Microscope, compound light

Focus Question

Design your own **Focus Question** based on the information presented in the Introduction and Materials sections above, as well as the additional information presented in the Procedure section that follows. Write your Focus Question on the Vee Form.

Knowing Side

With the help of your teacher, your class will discuss the **Subject Area** background for this investigation. Then, with the help of your classmates, list the **Concepts** and the new **Vocabulary** words on your Vee Form. In the **Concept Statements** section of your Vee Form, use these words in sentences that define and explain them.

Procedure

You will probably want to do step 1 before class using a nearby pond or lake.

1. Pour pond water into a culture dish until it is about two-thirds full. Include some decaying plant debris and mud from the bottom of the pond. Allow the sample to settle for 5 to 10 minutes.

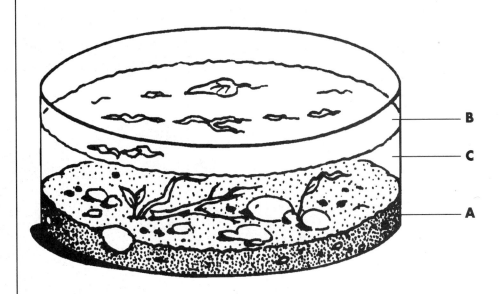

Vorticella

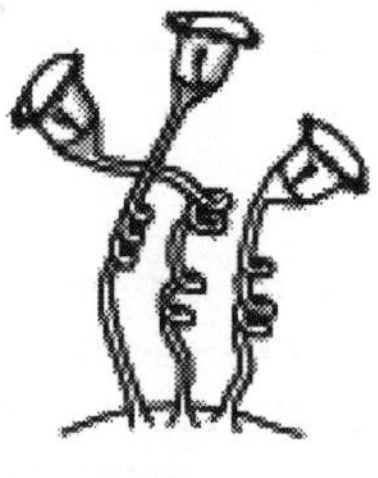

Paramecium

Roundworm

Cyclops

Fragilaria

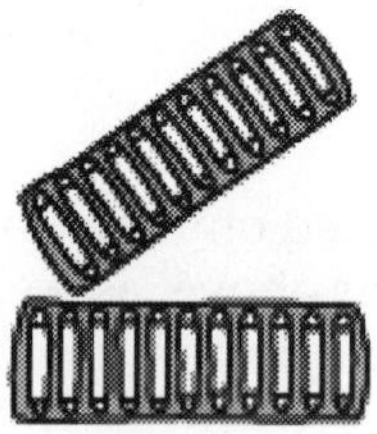

Halteria

Euglena

Stylonychia

Daphnia

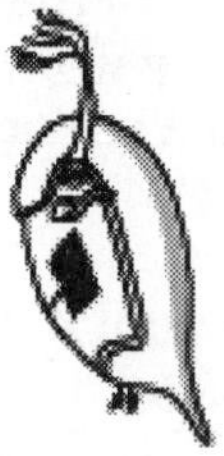

Stentor

Volvox

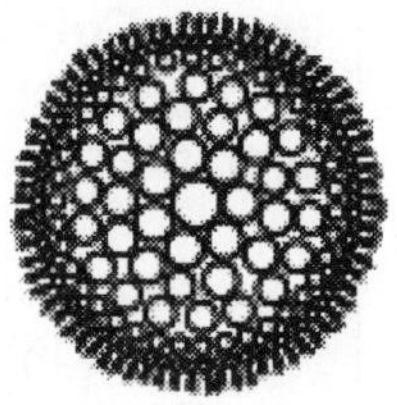

Chlamydomonas

Actinophrys

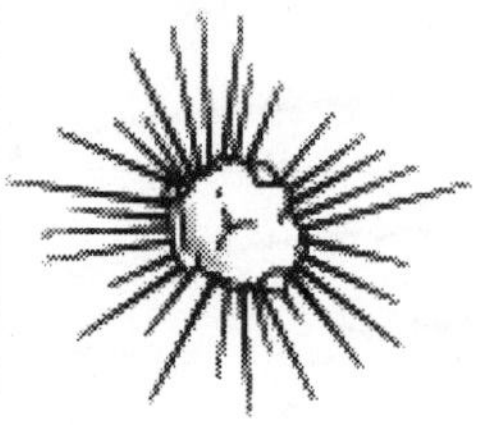

Colpidium

Phacus

Coleps

Blepharisma

Amoeba

Rotifer

Synura

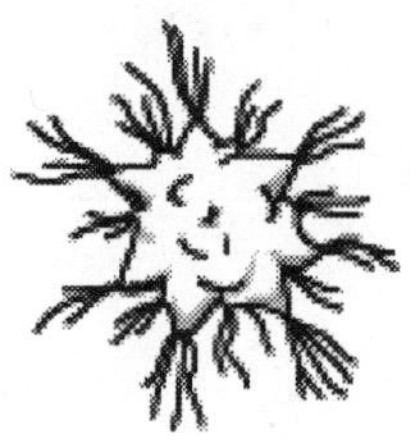

2. Label three slides A, B, and C, respectively. Prepare three wet mounts by taking samples from three different levels of the pond water: slide A, the bottom; slide B, the surface; slide C, approximately in the middle.
3. Observe each slide under the microscope. Use low power first, then switch to high power. Use the diagrams on the previous page to help you identify the organisms.
4. In the table found in the **Records** section of your Vee Form, list the organisms identified at each level of the pond water. Also describe any behaviors you observe, such as type of movement, feeding habits, or avoidance of obstacles. In the **Additional Records and Observations** section on the back of your Vee, make a sketch of each type of organism you find.
5. Compare your findings with those of other students in the class.

How can you account for the presence or absence of certain organisms on the slides of other students?

Water was taken from different levels of the pond water. Parts of the culture that contained a great deal of debris probably had a greater variety of organisms.

6. Clean up your materials and wash your hands before leaving the lab.

Use the information on the **Knowing Side** of the Vee to interpret your results from the **Doing Side**, and then write your **Knowledge Claim**. Write a **Value Claim** for this study.

Analysis

It is best to do this investigation in the fall or spring when the weather is cool but not cold.

When collecting the sample of pond water, be sure that samples are taken from different depths and locations. Be sure that they include some debris normally found in the pond.

1. Which slide had the greatest number of organisms? Explain.

Samples taken from debris near the bottom will have more organisms because they contain a source of food for the organisms.

2. After several weeks, what changes would you expect to take place in the culture dish?

Waste production accumulation, oxygen depletion, and increasing food scarcity will become evident. Therefore, some species will disappear, while others will increase and eventually become the dominant species.

Going Further

In the appropriate space on your Vee Form, write a **New Focus Question** that could be the point of a new investigation based on what you have just learned.

- For the next 2 weeks, examine the organisms in your pond water sample daily. Note any changes in the type and number of organisms.
- Take samples of pond water from the same location, but at different times during the year. Examine the samples and compare the number and types of organisms with those observed in this investigation. Write a brief hypothesis to account for changes of populations in pond water at different times of the year.

Name _____ Date _____ Class _____

Observing Pond Water

KNOWING SIDE

DOING SIDE

Focus Question:

What different life forms can be seen in pond water when viewed under a microscope? Is there a difference observed when samples are taken from different depths?

New Focus Question:

Would the organisms found in different levels be the same in a pond water sample taken at a different season?

Subject Area:

Microscopy
Ecosystems

Concepts:

microscopic life
ecosystem

Vocabulary:

protozoa
invertebrates

Concept Statements:

1. Pond water contains many protozoan organisms.
2. Pond water represents part of an aquatic ecosystem.

Value Claim:

Samples of healthy pond water should show varied life forms.

Knowledge Claim:

The debris of the culture contained a greater variety of organisms because it is probably a source of foods. Some organisms moved, some were attached to the debris.

Records:

Levels	Organisms Identified	Behavior Observed
1		
2		
3		

Materials:

Pond water, culture dish, wax pencil, slides (3), coverslips (3), medicine dropper, compound microscope.

Procedures:

Observe 3 samples of pond water taken at different depths on 3 wet mounts. Identify the organisms seen and describe their behaviors.

Additional Records and Observations:

Prokaryotic and Eukaryotic Algae

Introduction

In Chapter 3 of your textbook, the differences between prokaryotic and eukaryotic cells are described. Prokaryotic cells do not have a nucleus. They do not have true chromosomes, but simply chromatin material scattered throughout the cell. All bacteria, including cyanobacteria, are prokaryotic cells. Eukaryotic cells are those that have a true nucleus enclosed in a nuclear membrane. These cells have true chromosomes.

Another key difference is that the prokaryotic cells of cyanobacteria (blue-green algae) do not store the food they make during photosynthesis as starch, and cells of photosynthetic eukaryotes do. True starches turn blue-black in the presence of iodine. This fact can be used to determine whether a photosynthetic cell is prokaryotic or eukaryotic. In this investigation, you will do just that with three different types of algae.

Materials

- Starch suspension (in dropper bottle)
- Iodine solution (in dropper bottle)
- Spot plate
- Microscope, compound light
- Slides (3)
- Coverslips (3)
- Lens paper
- Wax pencil
- Algae samples (marked A, B, C)
- Watch glasses (3)

Focus Question

Design your own **Focus Question** based on the information presented in the Introduction and Materials sections above, as well as the additional information presented in the Procedure section that follows. Write your Focus Question on the Vee Form.

Knowing Side

With the help of your teacher, your class will discuss the **Subject Area** background for this investigation. Then, with the help of your classmates, list the **Concepts** and the new **Vocabulary** words on your Vee Form. In the **Concept Statements** section of your Vee Form, use these words in sentences that define and explain them.

Procedure

Suggested algae: cyanobacteria, such as *Anabaena, Nostoc, Oscillatoria*; green algae, such as *Chlorella, Spirogyra, Ulothrix*

It is suggested that you purchase and distribute to your students pure strains of algae. You may want to use two samples of green algae and one of cyanobacteria, or one of green algae and two of cyanobacteria. See the annotation following Going Further for Ward's catalog numbers of the suggested algae.

1. **CAUTION: Iodine is a skin irritant. Wear goggles, gloves and lab apron.**
2. Place 2 or 3 drops of starch suspension on a spot plate.
3. Cover the starch suspension with 2 or 3 drops of iodine solution. Observe and record your observations in the table provided in Table 1 of the **Records** section of your Vee Form.
4. Make a wet mount of each sample of algae (A, B, C). Observe each under the microscope, first under low power and then under high power. On the back of the Vee Form in the **Additional Records and Observations** section, make accurate drawings of what you observe.
5. Label the watch glasses A, B, and C. On each watch glass, put a 2-mL sample of the corresponding alga, A, B, or C.
6. On each watch glass, place 2 mL of iodine solution. Observe. Record your observations in Table 1.
7. Clean up your materials and wash your hands before leaving the lab.

Use the information on the **Knowing Side** of the Vee Form to interpret your results from the **Doing Side**, and then write your **Knowledge Claim**. Write a **Value Claim** for this study.

Analysis

1. From the drawings you made, which sample(s) showed a true nucleus bounded by a nuclear membrane?

   the samples of green algae, but not in any of the cyanobacteria

2. Is there any advantage to a true nucleus bounded by a nuclear membrane?

   A nuclear membrane has the advantage of being able to control the substances taken into the cell nucleus.

3. Which sample or samples of algae were prokaryotes?

   Answers will vary depending on the type of alga chosen.

4. Which sample or samples of algae were eukaryotes?

   Answers will vary depending on the type of alga chosen.

Going Further

In the appropriate space on your Vee Form, write a **New Focus Question** that could be the point of a new investigation based on what you have just learned.

The following suggested algae may be purchased from Ward's.

Cyanobacteria	Green Algae
Anabaena (86 M 1800)	**Chlorella** (86 M 0126)
Nostoc (86 M 2150)	**Spirogyra** (86 M 0650)
Oscillatoria (86 M 2300)	**Ulothrix** (86 M 0750)

Name _____ Date _____ Class _____

Prokaryotic and Eukaryotic Algae Vee Form Report Sheet

DOING SIDE

Value Claim:
Stains can be used to identify unknown microorganisms.

Knowledge Claim:
Eukaryotic algae store food as starch, which turns blue-black with iodine stain.

Records:
Table 1 Iodine Test for Cell Structure

Sample	Color
Starch + iodine	
Watchglass A	
Watchglass B	
Watchglass C	

Materials:
dropper bottles of starch suspension and iodine solution, spot plate, microscope, 3 slides and coverslips, lens paper, wax pencil, samples of algae (marked A, B, C), watch glass

Procedure:
Examine algae under a microscope and then test for the presence of starch. Identify the eukaryotes.

KNOWING SIDE

Focus Question:
How can prokaryotic and eukaryotic cells from three algae samples be distinguished under a microscope?

New Focus Question:
If cyanobacteria are prokaryotes, do they store food?

Subject Area:
Cell theory

Concepts:
staining

Vocabulary:
prokaryote
eukaryote

Concept Statements:
1. Prokaryotic cells are primitive and have no membran-bound structures.
2. Eukaryotic cells have nuclei and membrane-bound organelles.
3. Starch and iodine combine to give a blue-black color.

Additional Records and Observations:

	Alga A	Alga B	Alga C
Low Power			
High Power			

Plant Growth and Soil Conditions

Introduction

Plants growing in a mineral-deficient environment not only grow poorly, but also show various signs of the deficiency in their physical appearance. A trained observer can predict the mineral deficiency of a given soil by examining the plants growing in it, since each deficiency consistently displays its characteristic symptoms. If the leaves turn yellow, this is diagnosed as chlorosis. The mineral prescribed to promote healthy growth may depend on the age of the leaf and what part of the leaf shows signs of chlorosis. A plant diagnosed with necrosis may have leaves with dead, brown spots, and the placement of these dead areas will indicate which mineral the soil lacks.

In this investigation, you will observe the appearances of plants grown under various mineral-deficient conditions, using 10-day-old seedlings that have been grown in a root environment such as vermiculite or clean washed sand.

Materials

- Complete nutrient solution
- Mineral-deficient solutions (8)
- Sunflower seeds (1 packet)
- Tomato seeds (1 packet)
- Germination tray
- Distilled water
- Vermiculite or sand
- Cotton
- Disposable containers, 1 quart for nutrient solutions (10)
- Holders/lids for containers (10)
- Cork stoppers (10)

Focus Question

Design your own **Focus Question** based on the information presented in the Introduction and Materials sections above, as well as the additional information presented in the Procedure section that follows. Write your Focus Question on the Vee Form.

Knowing Side

With the help of your teacher, your class will discuss the **Subject Area** background for this investigation. Then, with the help of your classmates, list the **Concepts** and the new **Vocabulary** words on your Vee Form. In the **Concept Statements** section of your Vee Form, use these words in sentences that define and explain them.

Procedure

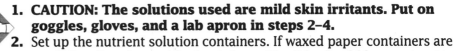

1. **CAUTION: The solutions used are mild skin irritants. Put on goggles, gloves, and a lab apron in steps 2–4.**
2. Set up the nutrient solution containers. If waxed paper containers are used, first line them with plastic bags. The lids should have cut-out centers, to allow the roots to be immersed in the solution when the plants are inserted into the holders. Use four-hole plant holders that will keep the plants upright after the holders are placed on top of the containers. To identify individual plants, use an appropriate marker and number the holes on each lid before the plants are inserted; keep a record of these in the **Additional Records and Observations** section on the back of the Vee Form. Fill the containers with the following solutions prepared by your teacher, then label each container according to both nutrient solution and seedling type:
   a. complete nutrient solution
   b. calcium-deficient solution

     c. sulfur-deficient solution
     d. magnesium-deficient solution
     e. potassium-deficient solution
     f. nitrogen-deficient solution
     g. phosphorous-deficient solution
     h. iron-deficient solution
     i. microelement-deficient solution
     j. distilled water

2. Remove a seedling carefully from the sand and wash the roots in distilled water to remove debris. Take a strand of cotton and wrap it firmly about the middle of the stem, as shown in Figure 1. Use enough cotton to support the plant when it is placed in the plant holder.

Figure 1

3. Insert three seedlings of the same type into the holder of each container, one plant per hole. When inserting the plants in the lid, be certain that the root system is far enough into the solution to be well immersed (see Figure 2). The roots must always be immersed; this must be checked daily. The cork stoppers are to be placed in hole #4 on the plant holder lids to serve as a convenient means for checking the level of the solution, and for the addition of distilled water.

Figure 2

4. Place the plants where they will receive the maximum available light. If they are on a windowsill, be sure to turn the containers a quarter turn every day. This prevents a bending to the light, which produces crooked plants.

5. After the plants are in place, height measurements should be made of each plant. All the plants should be measured at least every other day. Record these observations in Table 1 of the **Additional Records and Observations** section on the back of the Vee Form. Also note their general appearance, including the growth of the root system. After two weeks, deficiency symptoms should be apparent in all of the plants except the control (which received the complete nutrient solution). Continue the experiment for four weeks, if possible.

6. Record a summary of your observations in Table 2 on the back of your Vee Form.

7. Clean up your materials and wash your hands before leaving the lab.

Use the information on the **Knowing Side** of the Vee to interpret your results from the **Doing Side**, and then write your **Knowledge Claim**. Write a **Value Claim** for this study.

Analysis

1. What characterizes mineral deficiency in a plant?

some deficiency in plant growth caused by the lack of an important mineral for growth

2. Why must we wait two to four weeks to see observable results?

It may take two to four weeks for a mineral deficiency to become evident.

3. What was the control in this investigation? Why was one needed?

The control was the plant in the complete nutrient solution. It was needed as a contrast to the mineral-deficient plants.

4. Why is it important to observe the root system as well as the appearance of the leaves and stems?

Some minerals are crucial to root development, but may not have much effect on the leaves and stems.

5. How could a person well trained in mineral deficiencies in plants be valuable to the world of agriculture?

Such an expert could diagnose soil mineral deficiencies and be able to prescribe a fertilizer with the correct balance of minerals to ensure proper growth.

Going Further

In the appropriate space on your Vee Form, write a **New Focus Question** that could be the point of a new investigation based on what you have just learned.

- A number of interesting variations on the basic experiment are possible, including the use of weed seedlings that seem to grow in very unlikely and inhospitable places, such as cracks in a sidewalk.
- Vary the pH of the solutions to see if that has any effect on roots or leaves.

Place the seeds for the plants to be used in the experiment in clean sand or vermiculite. After planting, keep the rooting medium continually moist. Ten days before the experiments are scheduled, the seeds should be planted. Remembering that this growth takes two to four weeks to culminate in observable results, you may want to have your students set this investigation up well in advance of the date when the results will be relevant to the class material under study.

When the seedlings are of a sufficient size, they are transferred to the experimental growth media, which may be prepared from the stock solutions in a kit from a biological supply company. You may wish to order the following stock solution kit from Ward's.

Plant Mineral Deficiency Activity
(20 M 8400)

When the plant growth containers are filled and the lids in place, the plants are then selected. Allow your students to set up the growth containers in teams. From your seedlings choose 30 plants of as nearly equal size and appearance as possible. Give half of the class one seedling type and the other half the other seedling type.

Symptoms should be evident in two weeks, but it is best to maintain the experiment for a month.

Plant Growth and Soil Conditions

Investigation 18-1
Vee Form Report Sheet

KNOWING SIDE

DOING SIDE

Focus Question:

How can an observer predict the mineral deficiency of a given soil by examining the plants growing in it?

Value Claim:

A trained mineral deficiency observer can be invaluable for solving agricultural problems.

Knowledge Claim:

By observing the growth of the roots, stems, and leaves of plant seedlings, and the differential coloring of the leaves, one can determine which, if any, deficiency is present.

New Focus Question:

Will we observe the same results with weed seedlings that seem to grow in inhospitable places?

Records:

[See Additional Records and Observations on the following page]

Subject Area:

Plant physiology
Environmental science

Concepts:

nutrients
prediction
growth
diagnosis
symptoms

Vocabulary:

deficiency
chlorosis
necrosis
vermiculite

Materials:

Complete nutrient solution, mineral-deficient solutions (8), sunflower and tomato seeds, germination tray, vermiculite, cotton, disposable containers for nutrient solutions (10), holders and lids for containers (10), cork stoppers (10), distilled water.

Procedure:

Place seedlings in different mineral-deficient solutions, and observe the effects on growth and other plant characteristics over 2 – 4 weeks time.

Concept Statements:

1. Plants growing under conditions of mineral deficiency not only grow poorly, but also show signs of sickness.
2. Nutrient solutions can be varied to contain different plant nutrients or lack them.
3. If one knows the symptoms associated with a certain mineral deficiency, the plant can be nourished with a prescribed fertilizer.
4. Plants are chlorotic when the leaves are yellow.
5. Necrotic areas are dead/blackened parts of the plant.

Table 1 Plant Growth in Different Mineral-Deficient Solutions

Solution	Init.	Day 2	Day 4	Day 6	Day 8	Day 10	Day 12	Day 14
a. Complete								
b. Calcium-d								
c. Sulfur-d								
d. Magnesium-d								
e. Potassium-d			Answers will vary.					
f. Nitrogen-d								
g. Phosphorous-d								
h. Iron-d								
i. Microelements-d								
j. Distilled water								

Table 2 Symptoms of Mineral Deficiency in Plants

Mineral Deficient	Symptoms
Calcium	First symptom usually deformation of younger leaves, then a disintegration of terminal growing areas.
Iron	Young leaves light green or almost white, while older leaves are green. Unlike most other elements, iron cannot be withdrawn from the older leaves, so they retain a normal appearance. The yellowing or chlorosis of the younger leaves is most obvious in the intervenal areas.
Magnesium	Lower leaves show symptoms first, yellowing from the tip and eventually falling. Veins remain green longer than the intervenal areas.
Nitrogen	Leaves change from a green-yellow at the top to yellow to brown. Dead leaves at the bottom. In many species a red or purpling occurs along the veins.
Phosphorous	Plant stunted; leaves dark green; occsionally production of anthocyanins causes a red or purple color. Dead areas develop on leaves, petioles, and (in older plants), fruits, causing some dropping.
Potassium	Limited growth, weak stems, and a yellow mottling of the leaves; ultimately necrotic areas on leaf tips and edges; a general over-all yellowing appearance of the leaves.
Sulfur	Yellowing of the younger leaves in the early stages; ultimately an overall pale green may dominate.
Microelements	Manganese: Leaves become spotted with dead areas and fall. Copper: Tips of young leaves wither; plant may wilt even when it is watered. Zinc: Yellowing of lower leaves at tips and margins; leaves deformed. Molybdenum: Required for nitrogen metabolism, so symptoms resemble those of nitrogen deficiency.

Additional Records and Observations:

Stomata and Transpiration Rates

Introduction

As plants moved to the land from their water environment, the tendency to dessicate (dry out) became a problem. The cuticle, a thick, waxy layer of tissue, evolved as an adaptation for preventing this drying. The cuticle prevents the exchange of oxygen and carbon dioxide with the environment. Since photosynthesis cannot take place without this exchange, small pores, called stomata, evolved to allow for the movement of gases into and out of the plant. During the past few centuries, plant stomatal densities have decreased as carbon dioxide levels have risen.

Most plants lose 90 percent of the water taken into their roots by the process of transpiration. While plants do lose some water from their stems, most of the water is lost through the stomata of the leaves.

In this investigation you will observe the stomata in both terrestrial plants and floating aquatic plants. You will do a comparative stomatal count on leaves of the same species by making tape replicas of stomata. You will also observe the transpiration rates of several types of leaves.

Materials

Part A

- Leaves from a terrestrial plant
- Leaves from a floating aquatic plant
- Scalpel
- Microscope
- Slides (2)
- Coverslips (2)
- Medicine dropper
- Transparent tape
- Clear nail polish

Part B

- Strips of cobalt chloride paper (16)
- Transparent tape
- Plants
- Paper clips (16)
- Plastic bags (16)
- Twist ties (16)
- watch with second hand or timer

Focus Question

Design your own **Focus Question** based on the information presented in the Introduction and Materials sections above, as well as the additional information presented in the Procedure section that follows. Write your Focus Question on the Vee Form.

Knowing Side

With the help of your teacher, your class will discuss the **Subject Area** background for this investigation. Then, with the help of your classmates, list the **Concepts** and the new **Vocabulary** words on your Vee Form. In the **Concept Statements** section of your Vee Form, use these words in sentences that define and explain them.

Part A: Stomatal Density of Leaves

Procedure

Students may have a difficult time removing the epidermis. You may wish to either provide some already prepared epidermis or use prepared slides.

1. Select a leaf from a terrestrial plant. With the lower epidermis facing upward, bend and then tear the leaf at an angle. This will reveal a portion of the thin, colorless, lower epidermis. If live plants are not available, substitute a prepared slide of lower leaf epidermis.

HRW material copyrighted under notice appearing earlier in this work.

For Part A

- The tape stomata replica technique is intended as a tool for real research, not just as a lab exercise. Your students might want to go on to gather baseline information about stomatal densities on certain plants and make comparisons with similar plants at different times of the year, etc. The implications for environmental pollution effects are great, since stomata densities are reduced as plants are subjected to stress. Alfalfa is an excellent globally distributed indicator species.

Remind students that nail polish is flammable. They should not expose it to heat, sparks, open flame or other ignition sources.

2. **CAUTION: Scalpels are very sharp. Use extreme care.** With the scalpel, cut off some colorless epidermis tissue and make a wet mount of it. Place the slide on the microscope stage and focus under low power. Locate the stomata, then switch to high power.

3. Draw the stomata, the guard cells, and a few cells of the lower epidermis in Figure 1 of the **Additional Records and Observations** section on the back of your Vee Form. Count the stomata seen in the field of vision under high power and record the number in Table 1 on the back of the Vee Form.

4. Repeat the procedure in steps 1 through 3, using the upper epidermis of the same leaf. Then repeat these steps with both the lower and upper epidermis of a leaf from the chosen floating aquatic plant.

5. Record the data of each lab group on the chalkboard and calculate the average number of stomata corresponding to each heading. Record this information in Table 1.

6. Collect different leaf samples from your teacher. They can be dried or still on the plant. Note each tree's identity and the number of its leaves you will test.

7. Paint a 1-x-2 cm oval of clear fingernail polish on the leaves, avoiding ribbed veins. Paint both the leaf tops and undersides.

8. When the nail polish is dry, peel it off by placing transparent tape on the polish still on the leaf, then lift off the specimen by pulling the tape. Place this along with three other sample replica specimens on the same slide.

9. Count all stomata in one field of view at 400X. You'll see 12 to 30 stomata in your field of view. Count three areas on each replica and record your total in Table 2 on the back of the Vee.

Were all the leaves from the same tree consistent in their stomatal counts?

They will be very similar, but not likely identical.

Cobalt chloride paper is blue when dry and pink when moist. It is very sensitive to any water present, so it is best not to do this investigation on a humid day. Demonstrate the use of cobalt paper to your students so that they will be able to identify a positive test. You may do this by spraying a light mist of water onto a dry piece of cobalt paper.

Caution students that cobalt chloride is poisonous; be sure to remind them not to ingest any. Have them wash their hands immediately after handling it.

Part B: Leaf Transpiration Rates

1. Prepare 16 test strips for moisture detection by covering one-half of the front and back of each piece of cobalt chloride paper with transparent tape.

Why is one half of the cobalt chloride paper covered with tape, while the other half remains uncovered?

Since water cannot penetrate the tape, you will have a basis for comparison.

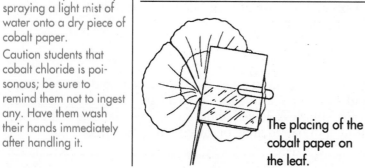

The placing of the cobalt paper on the leaf.

Enclosing leaf and cobalt paper in a watertight plastic bag.

These plants should be planted and growing unless otherwise indicated.

2. Place the test strip on a leaf that is still attached to the plant. Fasten the test strip to the leaf with a paper clip and enclose the leaf and test strip in a plastic bag with a twist tie. Note the amount of time that it takes for the test strip to turn pink while attached to the test leaf. Record this in Table 3 on the back of the Vee.

3. Repeat step 2 with another leaf on the same plant.

4. Repeat step 2 with the following pairs of leaves:
    a. stem of a plant and a leaf of the same plant
    b. large leaf and small leaf of the same plant.

Why is it important that the leaves come from the same plant in pairs a *and* b*?*

No factors should vary, other than the indicated one.

Are there other suggestions for choosing pairs of leaves that would make the data more reliable?

Answers will vary but could include testing or eliminating other possible variables.

5. Clean up your materials and wash your hands before leaving the lab.

Use the information on the **Knowing Side** of the Vee to interpret your results from the **Doing Side** for both parts of this investigation, and then write a **Knowledge Claim** for each. Write a **Value Claim** for this study.

Analysis

Part A: Stomatal Density of Leaves

1. Using information from the class data table, compare the number of stomata in the upper and lower epidermis of terrestrial plants.

   There are more stomata in the lower epidermis.

2. What advantage could this provide for the plant?

   It prevents the sun from drying out the plant when the stomata are open. It also prevents dirt from settling on the leaf and clogging up the stomata.

3. Using information from the class data table, compare the number of stomata in the upper and lower epidermis of aquatic plants.

   There are more stomata on the upper epidermis.

4. What advantage could this provide for the plant?

   Since the lower surface of the leaf is on the water, the upper surface should have more stomata to ensure adequate gas exchange.

5. Is the class information more reliable than the information gathered by one group?

yes, because more specimens were studied

Part B: Leaf Transpiration Rates

1. Which factor in each pair in step 4 had the greatest rate of transpiration?

There should be a greater rate of transpiration in the: leaf rather than stem; and in a large leaf rather than a small leaf.

2. What characteristics would be expected for a leaf that transpires at the most rapid rate?

living, smooth, broad, thin, or large

3. What characteristics would be expected for a leaf that transpires at the slowest rate?

fuzzy, needle-shaped, succulent, small, or dead

4. Which leaf type would be successful for a lawn?

Answers will vary but could include a small leaf which would reduce the need for water.

Going Further

In the appropriate space on your Vee Form, write a **New Focus Question** that could be the point of a new investigation based on what you have just learned.
- Compare the number of stomata found in plants that grow in dry regions to the number of stomata of those that grow in humid regions.
- A study in Israel using alfalfa leaves found seasonal differences in stomatal counts. Differences between irrigated and nonirrigated plant stomatal counts were also found. What might be done in your area using stomatal counts to generate valuable data about environmental effects?
- Set up a series of houseplants of the same type. Vary the amount of water given them, from none to an excessive amount. Study the rate of transpiration in each plant.

Name _____ Date _____ Class _____

Stomata and Transpiration Rates
Vee Form Report Sheet

DOING SIDE

Value Claim:
This can be a tool for "real" research about stomatal densities on specific plants. Variables could include specific growing conditions, specific times of year, and different pollution conditions.

Knowledge Claim:
A. Most stomata are on the upper surface of floating aquatic plants, while most stomata are on the lower surface of terrestrial plant leaves. Stomata density varies from leaf to leaf.

B. Greater transpiration occurs in thin, flat leaves, smooth-surfaced leaves, broad-surfaced leaves, and the lower surface of terrestrial leaves.

Records:
[See Additional Records and Observations on the following page]

Materials:
A. Transparent tape, clear nail polish, leaves from a terrestrial plant and aquatic plant, scalpel, microscope, 2 slides and coverslips, medicine droppers.

B. Cobalt chloride paper (16 strips), transparent tape, 16 paper clips, 16 plastic bags, twist ties, watch with second hand, plants.

Procedure:
A. Make longitudinal sections of plant leaf epidermal tissue from terrestrial and aquatic plants and compare number and location of stomata. Use transparent tape to make quick mounts of stomata replicas.

B. Compare transpiration rates in different kinds of leaves.

KNOWING SIDE

Focus Question:
A. How do stomata in plants from two different environments differ? Does stomata density remain constant from leaf to leaf?

B. How does leaf structure affect the rate of transpiration of water lost through the stomata?

New Focus Question:
Will stomata density be different from season to season, and is it the same as that of the same species growing in more polluted areas?

Subject Area:
Plant physiology
Environmental effects

Vocabulary:
dessicate
stomata
cuticle
transpiration

Concepts:
gas exchange
cobalt chloride

Concept Statements:
1. Dessication, or drying out, is a problem plants control by a cuticle covering that prevents gas exchange.
2. Stomata allow for gas exchange and transpiration of water through the leaves in plants.
3. Plants have shown decreasing stomatal densities in response to trends of increasing carbon dioxide concentrations.
4. Cobalt chloride paper is blue when dry and pink when moist.

Additional Records and Observations:

Table 1 Stomatal Densities

| Group | Terrestrial Plant | | Floating Aquatic Plant | |
	Lower Epidermis	Upper Epidermis	Lower Epidermis	Upper Epidermis
Class Average				

Figure 1

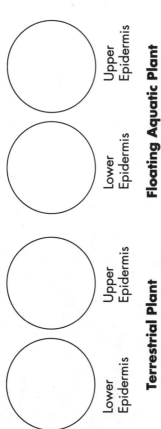

| Lower Epidermis | Upper Epidermis | Lower Epidermis | Upper Epidermis |

Terrestrial Plant   **Floating Aquatic Plant**

Table 2 Stomata Counts of Various Leaf Types

| Leaf Type | No. of Stomata in 3 Viewings of Leaf Replica | | | |
	View 1	View 2	View 3	Total
1.				
2.				
3.				
4.				
5.				

Table 3 Transpiration in Plants

Part or Type of Plant Tested	Time Required for Change
Leaf Attached to Plant	
Other Leaf Attached to Plant	
Stem	
Leaf	
Large Leaf	
Small Leaf	

Vitamin C From Foods

Introduction

Vitamin C is one of the important substances that contributes to the maintenance of normal health. For this reason, your diet should include foods rich in vitamin C, also known as ascorbic acid. Foods rich in vitamin C include fresh fruits, vegetables, and fruit juices. In this investigation, food products will be tested for vitamin C content, using indophenol, a blue indicator that becomes colorless in the presence of vitamin C.

Materials

In buying the juices for this investigation, avoid those labeled, "Vitamin C added." The vitamin C content found naturally in the juice is to be determined.

- Test tubes (11), 13 × 100 mm
- Indophenol solution, 0.1%
- Ascorbic acid solution, 0.1%
- Ascorbic acid solution, 0.05%
- Orange juice, fresh
- Orange juice, frozen
- Orange juice, canned
- Orange juice, boiled
- Orange juice, fresh, exposed to the air for one hour
- Lemon juice
- Pineapple juice
- Tomato juice
- Carrot juice
- Test tube rack

Focus Question

Design your own **Focus Question** based on the information presented in the Introduction and Materials sections above, as well as the additional information presented in the Procedure section that follows. Write your Focus Question on the Vee Form.

Knowing Side

With the help of your teacher, your class will discuss the **Subject Area** background for this investigation. Then, with the help of your classmates, list the **Concepts** and the new **Vocabulary** words on your Vee Form. In the **Concept Statements** section of your Vee Form, use these words in sentences that define and explain them.

Procedure

Indophenol and ascorbic acid can be obtained from Ward's. Place these solutions in dropper bottles for ease in dispensing.

Indophenol
(39 M 1670) 1g powder

Ascorbic Acid
(39 M 2005) 25g powder

Part A: Establishing the Standard

1. Place 15 drops of indophenol in one test tube. Add the 0.1% ascorbic acid solution, one drop at a time, to this test tube. Swirl the test tube after each drop. Continue until the indophenol becomes colorless.

How many drops of ascorbic acid solution are required to decolorize the indophenol solution?

Answers will vary.

2. Repeat step 1, using the 0.05% ascorbic acid solution.

How many drops of ascorbic acid are now required to decolorize the indophenol solution? Why?

The number should double. The ascorbic acid solution is half as strong as the previous one.

Show the students how to set up the proportion calculation as shown below. Example: If it takes 10 drops of 0.1% ascorbic acid to neutralize the indophenol, and 7 drops of an unknown concentration (x %), then:

$$\frac{10 \text{ drops}}{7 \text{ drops}} = \frac{x\% \text{ ascorbic acid}}{0.1\% \text{ ascorbic acid}}$$

Notice that this is an inverse proportion, since a larger number of drops corresponds to a weaker solution. Now complete the calculation as illustrated below.

$$(10)(0.1\%) = (7)(x\%)$$
$$1\% = 7(x\%)$$
$$\frac{1\%}{7} = (x\%)$$
$$0.14\% = x\%$$

Part B: Performing the Test

1. Place 15 drops of indophenol in a test tube. Slowly add fresh orange juice, one drop at a time, to the indophenol. Swirl the test tube after each drop. Continue until the indophenol becomes colorless. Record the number of drops of fresh orange juice required to decolorize the indophenol solution in Table 1 of the **Additional Records and Observations** section of your Vee Form.
2. Repeat step 1 for each of the following:
   - frozen orange juice
   - canned orange juice
   - boiled orange juice
   - orange juice exposed to air
   - lemon juice
   - pineapple juice
   - tomato juice
   - carrot juice
3. Using 0.1% ascorbic acid solution as the basis for comparison, calculate the percentage of ascorbic acid in each type of juice tested, and record this in your table. Set up the calculation as a proportion, as shown by your teacher.
4. Display your "Percentage of Vitamin C" data from Table 1 as a bar graph in Figure 1 on the back of the Vee Form.
5. Clean up your materials and wash your hands before leaving the lab.

Use the information on the **Knowing Side** of the Vee to interpret your results from the **Doing Side**, and then write your **Knowledge Claim**. Write a **Value Claim** for this study.

Analysis

1. Is this a control investigation? If so, what is the control?

   Yes; the ascorbic acid solution of known concentration is used as a standard.

2. What should be avoided to retain the natural vitamin C of juices?

   boiling and exposure to air

3. What juices can be included in the diet for high intake of vitamin C?

   Orange juice is richest in vitamin C, followed by lemon, pineapple, tomato, and carrot juice.

4. From the results of this investigation, suggest storage procedures for orange juice after it has been removed from the freezer compartment.

   It should be stored in a refrigerator in a sealed container.

Going Further

In the appropriate space on your Vee Form, write a **New Focus Question** that could be the point of a new investigation based on what you have just learned.
- Test vegetables and fruits for their vitamin C content. Compare these food products with the juices tested in this investigation.
- Test products that are labeled "Vitamin C added." Compare these products with the juices tested in this investigation.

Name _____ Date _____ Class _____

Vitamin C From Foods

KNOWING SIDE

DOING SIDE

Focus Question:

Which juice types have the highest concentration of vitamin C?

Subject Area:

Biochemistry
Nutrition

New Focus Question:

Does fresh orange juice that is stored in the refrigerator in a closed container lose its vitamin C content over time?

Concepts:

indicator
health
vitamins

Vocabulary:

indophenol
ascorbic acid

Concept Statements:

1. Vitamin C is important to normal health maintenance.
2. Vitamin C is ascorbic acid, found in fresh fruits, vegetables, and fruit juices.
3. Indophenol is a blue chemical indicator that turns colorless in the presence of vitamin C.

Value Claim:

If tomatoes are left out on a salad bar for awhile, they will lose their vitamin C content.

Knowledge Claim:

Boiling and exposure to air reduces vitamin C content. Fresh orange juice has more vitamin C than frozen, canned, or boiled orange juice. Orange juice has more vitamin C than lemon, pineapple, tomato, or carrot juices.

Records:

[See Additional Records and Observations on the following page]

Materials:

test tubes (11), indophenol solution (0.1%), ascorbic acid solution (0.1% and 0.05%), orange juice (fresh, frozen, canned, boiled, and exposed to room temperature for one hour), assorted juices (lemon, pineapple, tomato, carrot), test tube rack

Procedure:

Determine the number of drops of ascorbic acid needed to turn indophenol colorless (to establish the standard), then test how much of various juices is needed to turn the indicator colorless. Calculate the percentage of vitamin C in each juice type, then plot this on a bar graph.

Figure 1

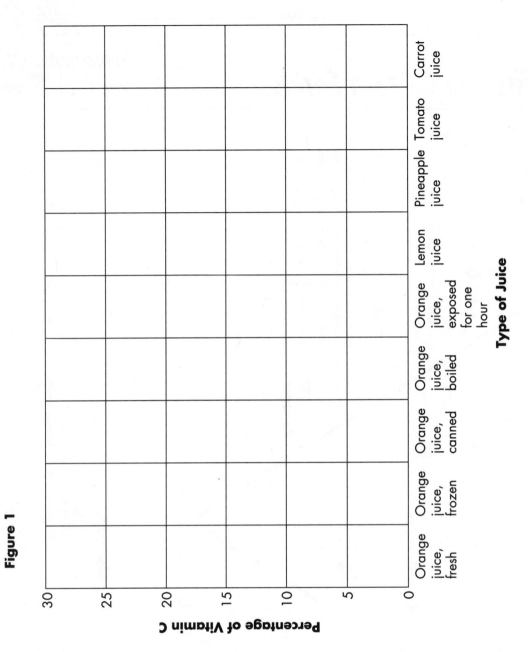

Percentage of Vitamin C

30 — 25 — 20 — 15 — 10 — 5 — 0

Orange juice, fresh | Orange juice, frozen | Orange juice, canned | Orange juice, boiled | Orange juice, exposed for one hour | Lemon juice | Pineapple juice | Tomato juice | Carrot juice

Type of Juice

Table 1

Juice	Drops of Ascorbic Acid	Percentage of Vitamin C
Orange juice, fresh		
Orange juice, frozen		
Orange juice, canned		
Orange juice, boiled		
Orange juice, exposed		
Lemon juice		
Pineapple juice		
Tomato juice		
Carrot juice		

Regulatory Chemicals of Plants

Introduction

Plant the bean seeds in sand 14 days before the lab is to begin.

The effects of gibberellic acid were first discovered in rice. "Foolish seedling disease" occurred in plants infected by a fungus whose gibberellic acid secretions caused the rice plants to become taller than normal, until they finally fell over. There are many growth-regulating substances in plants. In this investigation you will observe the effects of two of these, gibberellic acid and indoleacetic acid, on plant growth.

Materials

Caution students that IAA and GA are mild skin irritants, and should not be ingested.

- Flasks (5)
- Wax pencil
- Labels
- Indoleacetic acid solution (IAA), 1:10,000
- Indoleacetic acid solution (IAA), 1:1,000,000
- Gibberellic acid solution (GA), 1:1,000,000
- Bean seedlings (5)
- Scalpel
- Trowel
- Cotton
- Metric Ruler

Focus Question

Design your own **Focus Question** based on the information presented in the Introduction and Materials sections above, as well as the additional information presented in the Procedure section that follows. Write your Focus Question on the Vee Form.

Knowing Side

With the help of your teacher, your class will discuss the **Subject Area** background for this investigation. Then, with the help of your classmates, list the **Concepts** and the new **Vocabulary** words on your Vee Form. In the **Concept Statements** section of your Vee Form, use these words in sentences that define and explain them.

Procedure

The solutions used in this lab may be made from the following Ward's items.

3-Indoleacetic Acid (39 M 1660) 1g powder

Gibberellic Acid (39 M 1445) 1g crystal

Remind students not to harm the other plant parts as they remove the cotyledons.

1. Label five flasks as follows:

   flask 1: #1-water
   flask 2: #2-IAA 1:10,000
   flask 3: #3-IAA 1:1,000,000

   flask 4: #4-water
   flask 5: #5-GA 1:1,000,000

   Put your initials on each flask, using a wax pencil.
2. Fill the flasks two-thirds full with the solutions that correspond to their labels.
3. **CAUTION: Scalpels are very sharp. Be careful not to cut yourself.** Using a scalpel, carefully cut the stems of three bean plants near the soil level. Also remove the cotyledons from the plants.
4. Place one plant each in flasks 1, 2, and 3. Make sure the stems are submerged in the liquid. Pack cotton at the top of the flasks to keep the plants upright. Be careful not to crush or break the stems when packing the cotton.

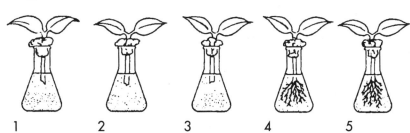

1   2   3   4   5

5. Set the three flasks aside in a place where the plants will receive light.
6. Observe the plants daily for 10 days. Use Table A in the **Additional Records and Observations** section on the back of your Vee Form to record the number of roots that appear, as well as their location.
7. Use a trowel to carefully remove the remaining two plants from the sand. Take a large clump of sand with the roots. Carefully shake the sand back into its container and rinse the roots gently in water to remove the remaining sand.
8. Remove the cotyledons from the stems of both plants with a scalpel.
9. Place one plant each in flasks 4 and 5, making sure the roots nearly touch the bottom of the flask.
10. Pack cotton carefully around the stem at the top of the bottle. The plants will then remain upright. Be careful not to crush or break the stem when packing the cotton.
11. Place these flasks with the others.
12. Measure the height of each plant for 10 days. Measure from the first pair of leaves to the tip of the stem, and record the data in Table B on the back of your Vee Form.
13. Clean up your materials and wash your hands before leaving the lab.

A tablespoon may be used to uproot the plants, if desired.

Use the information on the **Knowing Side** of your Vee Form to interpret your results from the **Doing Side**, and then determine your **Knowledge Claim**. Write a **Value Claim** for this study.

Analysis

Gibberellins are also involved in flowering, seed germination, and the breaking of seed dormancy. Gibberellins interact with other hormones. Different plants produce their own types of gibberellins, yet a species is sensitive only to the gibberellins it produces, although the effects of various gibberellins may be similar. After this lab investigation, you might go on to show that GA has a greater effect when it is placed on the most actively growing area of the plant.

1. How does gibberellic acid (GA) influence growth of bean plants?

   Bean stem and leaf growth is increased when gibberellic acid is present.

   _____

   _____

2. Does the strength of IAA affect root growth? Explain.

   Yes; indoleacetic acid, when present in small amounts (1:1,000,000), stimulates the growth of new roots. When IAA is present in large amounts (1:10,000), it inhibits root growth.

   _____

   _____

3. How might the knowledge of plant regulatory chemicals be of use to farmers and other plant growers?

   Farmers might use plant regulatory chemicals to treat deviations in normal growth patterns of crops, or to stimulate crop growth.

   _____

   _____

Going Further

In the appropriate space on your Vee Form, write a **New Focus Question** that could be the point of a new investigation based on what you have just learned.

Name _____ Date _____ Class _____

Regulatory Chemicals of Plants

KNOWING SIDE

Subject Area:

Homeostasis
Plant physiology, growth, and development

Focus Question:

What are the influences of indoleacetic acid and gibberellic acid on plant growth?

Concepts:

homeostasis
regulatory chemicals
hormones

Vocabulary:

indoleacetic acid
gibberellic acid

Concept Statements:

1. Plants respond to regulatory chemicals to maintain homeostasis (a balanced state).
2. Indoleacetic acid (IAA) and gibberellic acid (GA) are two plant regulatory chemicals for growth.
3. All organisms strive to maintain homeostasis.

DOING SIDE

Value Claim:

Because GA causes stems to elongate, it can be given to dwarf or bushy plants to make them taller.

Knowledge Claim:

IAA, when present in small amounts (1:1,000,000), stimulates the growth of new roots. When IAA is present in large amounts (1:10,000), it stops root growth. Stem elongation, as well as leaf growth, increases with GA (1:1,000,000).

New Focus Question:

Would varying the location of the application of regulatory chemicals affect the plant growth response?

Records:

[See Additional Records and Observations on the following page]

Materials:

flasks (5), wax pencil, labels, IAA (1:10,000), IAA (1:1,000,000), GA (1:1,000,000), bean seedlings (5), scalpel, trowel, cotton, metric ruler

Procedure:

Set up five flasks with bean seedlings to test the effects of IAA and GA regulatory chemicals over 10 days. Record results and compare observations to control flasks containing water.

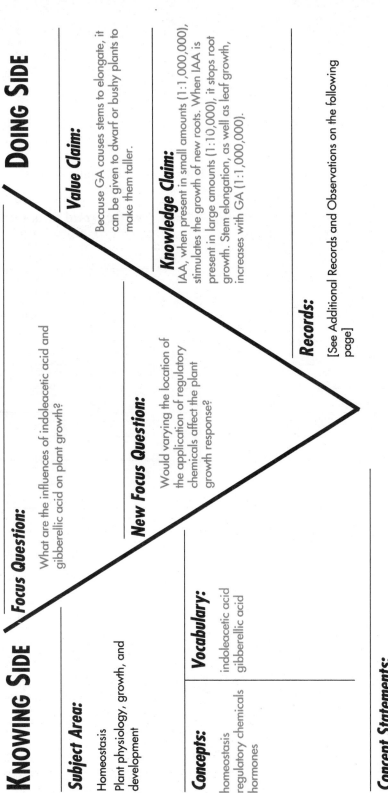

Table A Effects of Indoleacetic Acid on Plant Growth

Flask	Days	Roots	Location of Roots
#1 Water	0		
	1		
	2		
	3		
	4		
	5		
	6		
	7		
	8		
	9		
	10		
#2 IAA (1:10,000)	0		
	1		
	2		
	3		
	4		
	5		
	6		
	7		
	8		
	9		
	10		
#3 IAA (1:1,000,000)	0		
	1		
	2		
	3		
	4		
	5		
	6		
	7		
	8		
	9		
	10		

Table B Effects of Gibberellic Acid on Plant Growth

Flask	Days	Height	Total Change in Height (mm)
#4 Water	0		
	1		
	2		
	3		
	4		
	5		
	6		
	7		
	8		
	9		
	10		
#5 Gibberellic Acid (1:1,000,000)	0		
	1		
	2		
	3		
	4		
	5		
	6		
	7		
	8		
	9		
	10		

Additional Records and Observations:

Roundworms and Earthworms

Introduction

The vinegar eel is a roundworm. The complexity of roundworms, also known as nematodes, falls between that of flatworms and segmented worms. The roundworm has a primitive body cavity called the pseudo-coelom. Its digestive system consists of a tube with two openings, and it has an open circulatory system in which blood is pumped into open spaces in its body.

The earthworm is a segmented worm, or annelid. Some of the special structures of the earthworm are a segmented body, a coelom (a more elaborate body cavity in which organs are internally suspended by membranes), a complex nervous system, excretory organs, and a closed circulatory system in which blood travels through blood vessels. In this investigation you will examine the external anatomy of the vinegar eel, which is a nematode, and the earthworm, which is an annelid. You will then observe behavioral responses of earthworms to changes in their environment.

Materials

Part A	Parts B and C	
• Vinegar eels, live	• Earthworms, live	• Black paper
• Culture dish	• Beaker, 500 mL	• Tape
• Vinegar	• Paper towels	• Lamp
• Stereomicroscope	• Dissecting pan	• Ice, crushed
• Microscope, compound light	• Water, cold tap	• Water, hot tap
• Medicine dropper	• Hand lens	
• Depression slides		
• Coverslips		

Focus Question

Design your own **Focus Question** based on the information presented in the Introduction and Materials sections above, as well as the additional information presented in the Procedure section that follows. Write your Focus Question on the Vee Form.

Knowing Side

With the help of your teacher, your class will discuss the **Subject Area** background for this investigation. Then, with the help of your classmates, list the **Concepts** and the new **Vocabulary** words on your Vee Form. In the **Concept Statements** section of your Vee Form, use these words in sentences that define and explain them.

Procedure

Part A: Vinegar Eels

1. Collect some vinegar eels with a medicine dropper from the bottom of the container in which they are kept, then transfer them to your culture dish. Observe the vinegar eels by viewing the culture dish, using the low-power setting of the stereomicroscope, or by examining the contents of the dish with a hand lens.

Describe what you see in the culture dish under the scope.

vinegar eels thrashing back and forth from side to side

Vinegar Eel

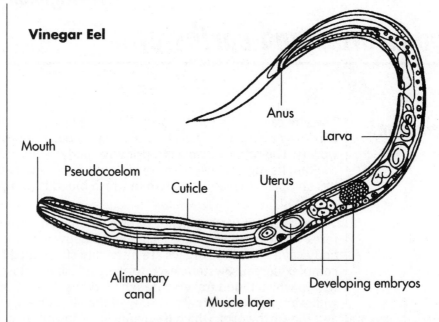

2. Make a vinegar eel wet mount by transferring a drop of the vinegar eel culture to a depression slide. You may improve your chances of capturing one or more vinegar eels in the drop of vinegar by gently swirling the vinegar in the culture dish and tilting the dish slightly. Cover the drop of vinegar with a coverslip and examine your slide using a compound light microscope set at low power.

3. Roundworms are generally rounded at the anterior (front) end and pointed at the posterior (back) end. Their bodies are covered with a tough cuticle. Make a drawing of the vinegar eel in the **Records** section of your Vee Form. Label the anterior end, the posterior end, the cuticle, the mouth, and the anus.

4. Roundworms have two sets of longitudinal (lengthwise placed) muscles on their dorsal (back) sides and two sets on their ventral (belly) sides. When the dorsal sets of muscles contract, the ends of the roundworm bend upward. When the ventral sets of muscles contract, the ends of the roundworm bend downward. Because roundworms do not have ring-like circular muscles, they cannot move from side to side. Observe the motion of a single vinegar eel on your slide. If it is still, gently tap the edge of the microscope slide with the end of your pencil.

Describe how the vinegar eel moves its body.

It contracts and extends its longitudinal muscles.

Is the vinegar eel able to move efficiently from place to place? Explain.

No, its thrashing motion allows no consistent direction of movement.

Focus on any vinegar eels that lie on the bottom of the depression slide.

Are these vinegar eels on the bottom of the depression slide able to move more efficiently than those closer to the surface? Explain.

Yes, they may push off the bottom and move as they drift in a specific direction.

In order to move from place to place, roundworms must be able to push their bodies against a solid surface, such as soil particles. Roundworms usually live in soil or on the bottoms of oceans, lakes, or streams.

5. Compare the vinegar eel you have seen with the diagram of the female vinegar eel shown on the previous page.

List the structures that you can identify in your wet mount.

Answers will vary; students should be able to see eggs.

Part B: Earthworms

1. **CAUTION: You are working with a live animal. Be sure to handle it gently and to follow all directions carefully.** Fold a paper towel and place it in the bottom of your beaker. Add just enough water to dampen the paper towel. Obtain an earthworm from your teacher and place it in your beaker.

2. Moisten another paper towel and place it in your dissecting pan. Place the earthworm on it. Frequently moisten the earthworm throughout this procedure by squirting tap water over it with a medicine dropper. Watch the earthworm move, and notice which end seems to lead. This is the anterior end. The thickened band around the body is called the clitellum.

3. In the **Additional Records and Observations** section on the back of your Vee Form, you will find a place to make your drawing of the earthworm. Label the anterior end, the posterior end, dorsal surface, ventral surface, and the clitellum.

4. Carefully pick up the earthworm and examine it. The mouth is located on the first segment, directly to the rear and underneath the prostomium, a small projection overhanging the first segment. Count the segments, beginning with the segment containing the mouth and ending with the segment containing the clitellum.

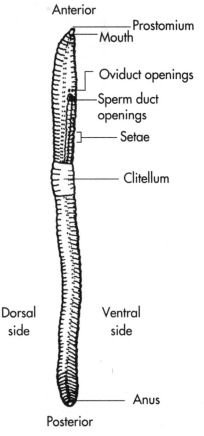

Anterior

Prostomium

Mouth

Oviduct openings

Sperm duct openings

Setae

Clitellum

Dorsal side

Ventral side

Anus

Posterior

How many segments are there before the clitellum?

32

5. Run your finger very lightly along the ventral side of the earthworm. The bristlelike structures are called setae, and have slightly curved, pointed ends. There are two pairs of setae on each segment. Each pair is attached to a set of muscles. Use a hand lens to examine the setae more closely.

What is the function of setae?

The setae provide traction as the earthworm moves through the soil.

Label the prostomium, mouth, setae, sperm duct, and oviducts on your drawing.

6. Gently lay the earthworm on its dorsal side.

How does the earthworm respond?

It rolls over to its ventral side.

7. Each segment of the earthworm is closed off from the segments in front of and behind it with a membrane. When the circular muscles around the segment are contracted, the fluid in the coelom of the segment is compressed and the segment becomes longer. When the circular muscles are relaxed and the longitudinal muscles that run lengthwise in the segment are contracted, the segment shortens. Watch your earthworm move. You may wish to use the hand lens to observe the segments more closely as it moves.

Describe how the earthworm travels.

The earthworm anchors the setae on its posterior segments and then extends the body in front of the anchored segments forward. It moves by stretching the anterior part of the body and pulling the posterior part of its body.

Part C: Earthworm Behavior

1. Place your earthworm back in the beaker. Then cover half of the dissecting pan with black paper and tape it firmly to the sides of the pan, as shown in the picture below. This will place half of the pan in the dark. Place three earthworms in the pan, so that the anterior ends of all the worms are in the light and the posterior ends are under the paper in the dark.

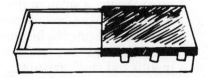

2. Place a lamp above the pan, so that light falls on the boundary between the black paper and the uncovered area for three minutes. In Table A of the **Additional Records and Observations** section on the back of your Vee Form, record the number of worms that moved toward the light and the number that moved toward the dark. Repeat the procedure three more times and record all results in Table A. Use new earthworms for each trial, if possible.

3. Repeat the procedure, but place the earthworms in the dissecting pan so that their anterior ends are in the dark and their posterior ends are in the light. Repeat the procedure three more times and record all results in Table A.

4. In the table, total each row and record under "Group totals." Put your totals on the chalkboard. Total all the groups' results under "Class totals."

5. Place the earthworms back in the beaker, and remove the black paper and paper towels from the dissecting tray. On one side of the dissecting tray, place a paper towel that has been soaked in crushed ice mixed with water. Next to the cold towel, place a paper towel soaked in hot tap water.

6. Place three earthworms in the dissecting pan, so that their anterior ends are on the cold towel and their posterior ends are on the warm towel. Wait three minutes, then record in Table B on the back of your Vee Form the number of worms that have moved toward the cold towel and the number that moved toward the warm towel.

7. Repeat the procedure, using fresh paper towels. Place the worms in the pan, so that their posterior ends are on the cold towel and their anterior ends are on the warm towel. Record your observations in Table B on your Vee Form. Repeat the procedure three more times and record all results in Table B on your Vee Form.

8. In Table B, total each row and record under "Group totals." Put your totals on the chalkboard. Total all the groups' results under "Class totals."

9. Place the earthworms back in the beaker, making sure that the paper towel in the bottom of the beaker is still damp. Remove the paper towels from the bottom of the dissecting pan and dry the bottom of the dissecting pan with a paper towel. Cover half of the bottom of the dissecting pan with a moistened paper towel and the other half with a dry paper towel. Place the earthworms back in the pan with their anterior ends on the dry paper towel and their posterior ends on the damp paper towel.

How do the earthworms respond?

Most earthworms will move toward the damp paper towel.

10. Clean up your materials and wash your hands before leaving the lab.

Use the information on the **Knowing Side** of the Vee to interpret your results from the **Doing Side**, and then write your **Knowledge Claim**. Write a **Value Claim** for this study.

Analysis

1. Contrast the movement of the vinegar eel with the movement of the earthworm. Explain why they move differently.

Vinegar eels thrash and whip their bodies because they do not have circular muscles.
Earthworms crawl forward with the help of setae and longitudinal, as well
as circular, muscles.

2. In addition to the membranes that separate the segments of the earthworm, each segment is also divided into right and left halves by another membrane. How do you think the earthworm turns to the right? To the left?

The earthworm turns to the right by contracting the longitudinal muscles on the right side
of the segment while relaxing muscles on the left side; it contracts longitudinal muscles on
the left while relaxing muscles on the right as it turns to the left.

3. Do the anterior and posterior ends of the earthworm respond equally to light? What evidence supports your answer?

No, the anterior end is more sensitive to light because most of the light-sensitive cells are
located there. When the posterior end was in the light, the worms did not tend to move.

4. How is the earthworm's response to light, temperature, and moisture beneficial to its survival?

The earthworm's usual movement toward darkness, moisture, and cooler temperatures
ensures that it stays in the soil and does not dry out.

Going Further

In the appropriate space on your Vee Form, write a **New Focus Question** that could be the point of a new investigation based on what you have just learned.

• Design an experiment that will show how earthworms respond to gravity. Write up the procedure, do the experiment, record observations, and state conclusions, according to what you have learned through your observations. Remember to have a control group for each of the variables you test.

Name _____ Date _____ Class _____

Roundworms and Earthworms

KNOWING SIDE

Subject Area:

Classification
Ethology (animal behavior)

Focus Question:

How do the external structures and the movements of roundworms and earthworms compare? How do earthworms respond to light, temperature, and moisture?

New Focus Question:

How do earthworms respond to gravity?

Concepts:

behavior
classification

Vocabulary:

annelid
nematode
cuticle
setae
clitellum
prostomium

Concept Statements:

1. Earthworms have more complex body structures than roundworms.
2. Ethology is the study of behavioral responses of animals to changes in their environment.
3. Earthworms are annelids, which means they are segmented.
4. Nematodes are roundworms that show intermediate complexity between flatworms and segmented worms in classification schemes.
5. The cuticle is a protective coating. Setae are bristles.
6. The clitellum is a thick band around the earthworm used in reproduction.
7. The prostomium is a liplike projection at the earthworm's anterior end.

DOING SIDE

Value Claim:

Anyone who fishes using earthworms for bait would want to use the behavioral information gathered to lessen the stress on the worms. The worms should be stored in dark, moist, cool soil.

Knowledge Claim:

Vinegar eels thrash and whip their bodies. Earthworms have circular band muscles and can crawl with the help of setae bristles. The earthworm's movement toward darkness, moisture, and cooler temperatures ensures that it stays in the soil and does not dry out.

Records:

Vinegar Eel Drawing

Materials:

A. live vinegar eels, culture dish, vinegar, stereomicroscope, compound light microscope, medicine dropper, depression slides, coverslips
B., C. live earthworms, beaker (500 mL), paper towels, dissecting pan, water, hand lens, black paper, tape, lamp, crushed ice

Procedure:

A. Observe vinegar eels under the stereomicroscope and describe their movements and external structures.
B., C. Examine the structures of the earthworm and observe its responses to light, cold, and moisture.

Additional Records and Observations:

Earthworm

Table A Earthworm Responses to Light

		Trial 1	Trial 2	Trial 3	Trial 4	Group totals	Class totals
Anterior half in light	Moved toward light						
	Moved toward dark						
Anterior half in dark	Moved toward light						
	Moved toward dark						

Table B Earthworm Responses to Temperature

		Trial 1	Trial 2	Trial 3	Trial 4	Group totals	Class totals
Anterior end on cold towel	Moved toward cold						
	Moved toward warmth						
Anterior end on warm towel	Moved toward cold						
	Moved toward warmth						

Insect Behavior and Bioluminescence

Introduction

Success in the world of living things depends on an organism's ability to live long enough to pass its genes on to the next generation. Insects survive and thrive because they respond to their environments in special ways. Because of the insect's body plan, its highly specialized body parts, and its many unique forms of behavior, the insect is the most abundant and successful type of animal on Earth. In this investigation you will make cages in which to put insects in order to study their behavior. You will collect insects and use a dichotomous key to identify them according to taxonomic order. The chemical reaction that occurs during firefly bioluminescence will then be demonstrated.

Materials

Screening must be pliable in order to make the cages. Nylon, vinyl, or plastic is suggested.

- Jars with air holes in lids
- Insects, living
- Screening (one yard)
- Circles of cloth
- Heavy thread
- Needle
- Soil
- Flower pot
- Grass
- Dark fabric (one yard)
- Firefly lantern powder
- ATP powder
- Distilled water

Focus Question

Design your own **Focus Question** based on the information presented in the Introduction and Materials sections above, as well as the additional information presented in the Procedure section that follows. Write your Focus Question on the Vee Form.

Knowing Side

With the help of your teacher, your class will discuss the **Subject Area** background for this investigation. Then, with the help of your classmates, list the **Concepts** and the new **Vocabulary** words on your Vee Form. In the **Concept Statements** section of your Vee Form, use these words in sentences that define and explain them.

Procedure

Making the insect cage and collection of insects can be done by the students outside of class.

Certain butterfly species are protected by law. Contact your regional office of the U.S. Fish and Wildlife Service to obtain a list of protected species in your area.

Part A: Insect Collection and Identification

1. Form the screening into a roll, making it about 0.5 cm smaller in diameter than the flower pot. Sew or staple together the overlapped edges.
2. By sewing the circle to the top edge of the screen cylinder, close the top of the cylinder with the circle of cloth.
3. Fill the flower pot with soil. Place a clump of grass on top of the soil.
4. Insert the screen structure in the pot, so that it snugly fits inside the pot. The cage is now ready to use.
5. Insects are easily found. Some particularly good places to look are areas where there is food, in and under rotting logs, under stones, on plants, and around lights at night. Indoors, try looking in basements and around windows.

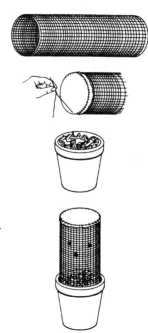

Many types of insects can sometimes be difficult to find due to their natural camouflage. Test your powers of observation by trying to spot them as they attempt to blend in with their surroundings by utilizing their natural colors or shapes. Many moths can be found on tree trunks, leafhoppers disguise themselves as "thorns" or branches, while walking sticks look like twigs. Can you find other examples?

Where else could insects be found?

garbage dumps, underbrush, in trees that secrete a sap, inside flowers
_____
_____

6. Collect the insects and place them in small jars with air holes in the tops. You might want to note what type of plant the insect was on, in what area you found it, whether it was in sun or shade, and what time of day or night it was found. Be sure to keep your live collected insects out of direct sunlight. In a closed container, heat caused by sunlight can kill them.

7. Read the dichotomous key provided on the next page.

You may wish to provide a special dichotomous key for insects that are common in your area.

What does dichotomous mean?

having only two opposing choices; selection of only one of two directions is possible
_____
_____
_____

8. To use a dichotomous key, follow these instructions. Read #1 of the key. Either 1a or 1b will describe the insect you are identifying. Follow the number given after your choice, 1a or 1b. For example, if the specimen in question was a bee, you would find it had four wings (1b), so you would proceed to #2 as instructed. You then observe that its wings are not membranous and covered with scales (2b), so you would proceed to #3. Then you note its hindwings are smaller than its forewings and its wings are veined (3a). The specimen belongs to the order Hymenoptera.

9. Classify, according to order, each insect that you have collected. Begin with #1 on the key for each specimen. Record the order name of each specimen in the **Additional Records and Observations** observation table on the back of your Vee Form.

Dichotomous Key to Major Orders of Insects

1a	Only two wings, hind wings have been reduced to knobby balancing organs, mouth parts adapted for sucking	**Diptera** (flies)
1b	Four wings	2
2a	Wings, membranous, and covered with scales	**Lepidoptera** (moths and butterflies)
2b	Not so	3
3a	Hindwings smaller than forewings, wings membranous and veined, or insect wingless and antlike in appearance	**Hymenoptera** (wasps, ants, and bees)
3b	Not so	4
4a	Top pair of wings horny, veinless, meeting in a straight line down center of back. Bottom wings membranous	**Coleoptera** (beetles)
4b	Not so	5
5a	Hindwings as large as or larger than forewings, wings finely netted, nodes on front edge of forewing	**Odonata** (dragonflies or damselflies)
5b	Not so	6
6a	Wings small, nearly absent. Large forceps on rear of abdomen	**Dermaptera** (earwigs)
6b	Not so, wings leathery	7
7a	First pair of wings thickened at base, shortened or membranous at tip. Scutellum, a scale-like portion of the upper thorax, is shield-shaped	**Hemiptera** (true bugs)
7b	Not so	8
8a	Wings held sloping at sides when at rest, mouthparts formed for piercing and sucking	**Homoptera** (cicadas, leafhoppers, scale insects)
8b	Not so	9
9a	Wings parchmentlike, with network of veins, hind legs suited for jumping	**Orthoptera** (grasshoppers, crickets, cockroaches)
9b	Winged members of this group have thin membranous wings, most are wingless and antlike in appearance, abdomen is broadly joined to the thorax, pale soft insects usually found in decaying wood	**Isoptera** (termites)

Part B: Observing Insect Behavior

1. Place the insects you have collected inside the cage, one at a time. It is naturally easier to observe insects in the lab, but remember that the behavior you see may not be "typical." The confined environment and close association with each other may alter their behavior. You may wish to make an observation of some insects and their behavior in the field under natural conditions. In either observation setting, be quiet and still, so as not to disturb the insects insofar as possible.

2. Change their environment by moving them from shade to sun. Observe and record their actions in the **Additional Records and Observations** section table on the back of your Vee Form. Remember that insects send out chemical messages, can make sounds with various parts of their bodies, and have various ritual movements for communication. Watch for any of these actions and record your observations in your Vee Form Records table.

3. Cover the cage to keep out all light with the dark fabric and then listen for any change in the sounds being made. Record any change in the table in your Vee.

4. Observe how the various insects respond to each other. Record your observations. When you are finished, release the collected insects.

Why should you release these insects?

Insects are generally beneficial, providing pollination and control of other types of insects, for

example.

Part C: Observing Bioluminescence

You may wish to do Part C as a demonstration before the class. For the demonstration of bioluminescence, the following kit is available from Ward's:

ATP Firefly Kit
(36 M 5402)

Your teacher has the materials to enable you to examine the phenomenon of bioluminescence.

1. **CAUTION: These materials are prepared with arsenic. Put on a lab apron, safety goggles, and disposable gloves.**

2. Add 3 mL of distilled water to the vial of powdered firefly lanterns. Replace the vial's dropper and cap and shake all contents gently for 1 minute.

3. Add 100 drops of distilled water to the ATP powder. Replace its dropper and cap and shake the vial gently for 1 minute.

4. Darken the room as much as possible for the greatest contrast.

5. Add 5 drops of the firefly lantern solution to a small, clean test tube.

6. Add 1 drop of the ATP solution to the test tube and observe the reaction.

What happens?

It lights up.

The production of light, or bioluminescence, occurs because of reactions that occur between luciferin and luciferase (present in the firefly lanterns) and the ATP and magnesium ions present in the ATP solution.

Why would bioluminescence be part of a good communication system in flying insects?

It can be seen from a distance.

7. Clean up your materials and wash your hands before leaving the lab.

Use the information on the **Knowing Side** of the Vee Form to interpret your results from the **Doing Side**, and then write your **Knowledge Claim.** Write a **Value Claim** for this study.

Analysis

1. What did you find to be the best location for insect collecting?

Answers will vary.

2. Why do you think certain insects stay in particular areas?

They live near their food source, or they have special adaptations for that environment.

3. What were some of the major differences in behavior observed among the various insects?

Answers will vary according to which insects were collected.

4. Make a list of insects that you observed and tell what food each one preferred, if you can.

Answers will vary.

5. How do environmental conditions affect insects' sounds?

Each insect responds differently to light, darkness, temperature, humidity, and other

conditions.

6. Did environmental changes affect other types of insect behavior, in addition to the sounds they made? Explain.

Answers will vary, but some reactions should change.

Going Further

In the appropriate space on your Vee Form, write a **New Focus Question** that could be the point of a new investigation based on what you have just learned.

- Using your text, write a short explanation of the two types of metamorphosis, complete and incomplete. Make a diagram of each type, or collect examples of the stages of both types for two different insects.

Insect Behavior and Bioluminescence

Investigation 22-1
Vee Form Report Sheet

KNOWING SIDE

Subject Area:

Classification
Animal behavior
Biochemistry

Focus Question:

1. How can an insect classification key be used to identify unknown insects?
2. How are specific insect behaviors related to their survival?

New Focus Question:

Are some of the apparently different insects collected really the same species, but in different stages of metamorphosis?

Concepts:

survival
behavior
observation
collection

Vocabulary:

dichotomous key
Diptera
Lepidoptera
Hymenoptera
Coleoptera
Odonata
Dermaptera
Hemiptera
Homoptera
Orthoptera
Isoptera
bioluminescence

Concept Statements:

1. Behavior can be observed and described.
2. Organisms exhibit qualities selected for survival over many years of evolution.
3. Classification enables organisms to be systematically identified by the use of a dichotomous key.
4. Bioluminescence is caused by a biochemical reaction within the organism.

DOING SIDE

Value Claim:

A study of insect behavior may reveal the role of insects in the ecosystems to which they belong.

Knowledge Claim:

Insect keys are either/or guides to identifying unknown collected insects. Bioluminescence and sound (both detected over a great distance), flight, and chemical transmission all enable insects that may be far apart to achieve direct communication for reproduction.

Records:

[See Additional Records and Observations on the following page]

Materials:

living insects, screening (one yard), circles of cloth, heavy thread, needle, soil, flower pot, grass, honey, dark fabric, firefly lantern powder, ATP powder, distilled water.

Procedure:

Make an insect cage for observation. Collect insects. Observe the behavior of collected insects in shade and sun, and record reactions. Listen for their sounds. Observe two or more insect types together. Observe bioluminescence.

Additional Records and Observations:

Insect Order	Sun	Shade	Sound	Other Behaviors
1.				
2.				
3.		Answers will vary		
4.				
5.				

Snails

Introduction

You can collect land snails from moist, cool areas in woods or gardens. Freshwater snails can be found in lakes, ponds, marshes, and quiet areas in streams. Both land and freshwater snails can also be ordered from Ward's.

Land Snails
(87 M 4306) Pkg. of 6
(87 M 4312) Pkg. of 12

Materials

Aquatic Snails
(87 M 4350) Set of 25

Focus Question

Knowing Side

Procedure

Maintain land snails in a terrarium that mimics the environment from which you collected them. If ordering land snails from a biological supply company, be sure to find out how to maintain them.

Mollusks are found in fresh water, in salt water, and on land. Their soft bodies are covered by a specialized tissue called the mantle. The mantle is composed of a hard, protective material that forms either an inner or outer shell. Mollusks with outer shells are either bivalves (having two shells), or univalves (having one shell). The largest class of mollusks is the gastropods. Snails are gastropods that have an outer shell; slugs lack this external protective covering.

Snails are found in all three mollusk environments. Aquatic snails obtain oxygen by means of gills. Terrestrial snails and slugs respire through blood vessels in the mantle cavity, which absorb oxygen more efficiently than gills do for aquatic snails. Because the blood tissues in the mantle cavity must be kept moist to allow for the exchange of gases by diffusion, most land snails are found in cool, damp environments. Both land snails and freshwater snails eat either plants or the remains of dead organisms.

- Terrarium, with land snails
- Petri plate, bottom half
- Pencil
- Shoe box, with lid cut in half
- Lamp
- Aquarium, freshwater with snails
- Hand lens
- Stereomicroscope

Design your own **Focus Question** based on the information presented in the Introduction and Materials sections above, as well as the additional information presented in the Procedure section that follows. Write your Focus Question on the Vee Form.

With the help of your teacher, your class will discuss the **Subject Area** background for this investigation. Then, with the help of your classmates, list the **Concepts** and the new **Vocabulary** words on your Vee Form. In the **Concept Statements** section of your Vee Form, use these words in sentences that define and explain them.

Part A: Land Snails

1. Observe the land snails in the class terrarium. Watch them move. Use a hand lens to observe how a snail moves its foot. Look for mucus trails on the glass sides of the terrarium. A special gland in the front of the snail's foot produces mucus that helps the snail move.

Land snail

Describe the method by which snails move.

Snails move by contracting their foot in a wavelike motion from back to front. They glide over a mucus trail laid down by the front of the foot.

2. **CAUTION: You are working with a live animal. Be sure to handle it gently and to follow all directions.**
3. Carefully remove a land snail from the terrarium and place it in the bottom half of a petri plate. Use a hand lens or a stereomicroscope to examine the features of the snail. Draw your snail and record any observations in the space provided in the **Additional Records and Observations** section on the back of the Vee Form. Label the mouth, radula, tentacles, eyes, shell, and foot.

How many tentacles does the land snail have?

two pairs

Where is the visceral mass located? Look up the word "visceral" in a dictionary, if necessary.

inside the snail's shell

Where is the mantle located?

between the visceral mass and the shell

4. Look down on the top, or apex, of the shell. Follow the direction of the coil from the apex to the end, or lip, of the shell.

Does the spiral coil clockwise or counterclockwise?

Answers will vary, although most students will observe shells that spiral clockwise.

In most instances, the snail's shell spirals clockwise, or to the right. These right-handed, or dextral, shells are the most common. Left-handed, or sinistral, shells spiral counterclockwise and are uncommon.

Dextral shell   Sinistral shell

5. Very gently touch one of the snail's tentacles with the eraser end of a pencil.

Describe how the snail responds.

The snail will probably pull its tentacles inside its head; it may even pull its whole body into its shell.

6. Shine a lamp over one end of a shoe box and cover the other end with the half lid. Carefully place your snail in the illuminated end of the box. Check your snail about 15 minutes later. Your teacher may ask you to go on to the next activity while you are waiting. Repeat the activity, but place the snail in the darkened end of the box. When the activity is complete, return the snail to the terrarium.

In which end of the box did you find the snail?

the darkened end of the box

How does a snail sense light and dark?

with eyes that are located at the end of its tentacles

Part B: Freshwater Snails

1. Observe the freshwater snails in the class aquarium. Use a hand lens to observe how the snails move.

Freshwater snail

Describe how the freshwater snails move.

They move along the bottom or sides of the aquarium by contracting their feet in a wavelike motion, or they may float.

2. Find a snail that is eating algae on the side of the aquarium. Use a hand lens to observe the snail as it eats.

How does the snail remove algae from the glass?

It uses its radula, a rough, tongue-like organ used by some mollusks to eat; the snail scrapes up loose bits of food and moves the food into its digestive tract.

3. **CAUTION: You are working with a live animal. Be sure to handle it gently and to follow all directions.**
4. Carefully remove a snail from the aquarium and place it in the bottom half of a petri plate or shallow dish. Make sure the snail is always completely covered with aquarium water. Use a hand lens or a stereomicroscope to examine the external features of the snail. Draw your snail and record any observations in the space provided on the back of your Vee Form. Label the radula, mouth, tentacles, eyes, shell, and foot.

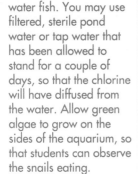

Maintain freshwater snails in an aquarium. Care for them as you would care for freshwater fish. You may use filtered, sterile pond water or tap water that has been allowed to stand for a couple of days, so that the chlorine will have diffused from the water. Allow green algae to grow on the sides of the aquarium, so that students can observe the snails eating.

How many tentacles does the freshwater snail have?

one pair

Where are the snail's eyes located?

Snails of the genus *Lymnaea* (commonly known as mud snails) and *Planorbis* (ramshorn

snails) have eyes located at the base of the tentacles.

5. Very gently touch one of the snail's tentacles with the eraser end of a pencil. After this, return the snail to the aquarium.

Describe how the snail responds when a tentacle is touched.

In most cases, the tentacles will probably not retract.

6. Clean up your materials and wash your hands before leaving the lab.

Use the information on the **Knowing Side** of the Vee to interpret your results from the **Doing Side**, and then write your **Knowledge Claim**. Write a **Value Claim** for this study.

Analysis

1. From your observations, describe the differences between land snails and freshwater snails.

Land snails have two pairs of retractable tentacles, and eyes located at the ends of one

pair of tentacles. Freshwater snails have one pair of tentacles that are not retractable, and eyes that are located at the bases of their tentacles.

2. On what basis are snails classified as mollusks?

Snails have soft bodies, a protective shell, and three distinct body parts: the head-foot, the

visceral mass, and the mantle.

3. What is the function of the mantle?

It secretes the shell.

Going Further

For freshwater snails, you may wish to use *Lymnaea* sp. or *Planorbis corneus* (ramshorn snails). These particular freshwater snails have lungs and must periodically go to the surface for air.

In the appropriate space on your Vee Form, write a **New Focus Question** that could be the point of a new investigation based on what you have just learned.

• Set up a freshwater aquarium or use the class aquarium to observe how freshwater snails reproduce. Observe the snails' mating behavior if possible, noting where they lay eggs, how long their eggs take to develop before hatching, and the appearance of the young snails.

Name _____ Date _____ Class _____

Snails

KNOWING SIDE

DOING SIDE

Focus Question:

How do land snails and water snails compare in their features and behaviors?

Value Claim:

Knowing the behavior of snails can help us protect the environment they need.

New Focus Question:

How do snails mate? Where do they lay eggs? How long does it take to hatch out young snails?

Knowledge Claim:

Snails are univalves. Snails move by contracting the foot in a wavelike motion from back to front. When a tentacle is touched, the land snail pulls its tentacles in and may pull its whole body into its shell. They prefer dark areas, and sense light with eyes at ends of tentacles (in land snails), and with eyes at base of tentacles (in aquatic snails). Snails scrape food with a radula in their mouths. Freshwater snails have only one pair of tentacles. Land snails have two pairs.

Records:

[See Additional Records and Observations on the following page]

Subject Area:

Classification
Mollusks

Concepts:

diffusion
terrestrial

Vocabulary:

mantle
bivalve
univalve
gastropod

Materials:

terrarium, land snails, petri plate (bottom half), pencil, shoe box with lid cut in half, freshwater aquarium with snails, lamp, hand lens, stereomicroscope

Procedure:

Observe land snails in a terrarium and examine foot movement with a hand lens, then draw observations. Test tentacle response to touch, and any preference for light or dark. Draw and observe freshwater snails in an aquarium, using a hand lens.

Concept Statements:

1. Mollusk soft bodies are covered by a mantle, which produces the hard inner or outer shell.
2. Bivalves have two shells; univalves have one shell.
3. Snails are in the largest class of mollusks, Gastropoda. They have outer shells, and live on land, in fresh water, or in salt water.
4. Most aquatic snails use gills to get oxygen from water by diffusion.
5. Land snails need to be moist for gas exchange, and live in cool, damp places.

Additional Records and Observations:

Land snail

Observations:

Freshwater snail

Observations:

Live Crickets

The class can share a single ice bucket for anesthetizing the crickets. Alternative methods for anesthetizing the crickets include placing them in the refrigerator for two or three hours before students examine them, or using a Ward's nonether anesthetic in the same manner as used for *Drosophila*.

Introduction

Polyethelene
Drosophila
Anesthetizer
(18 M 4950)

Crickets, grasshoppers, katydids, and locusts are members of the insect order Orthoptera. All orthopterans have two sets of wings and possess muscular hindlegs designed for jumping.

House crickets and field crickets are the most common species of crickets. House crickets, which were introduced into the United States from Europe, are slightly smaller than field crickets. Both house crickets and field crickets will chirp at any time of day or night, and rarely fly. In this investigation, you will observe cricket behavior, and locate and identify the external structures of the common house cricket or field cricket.

Materials

- Crickets, live
- Ice bucket
- Ice, crushed
- Beakers (3), 500 mL
- Apple pieces
- Hand lens or stereomicroscope
- Cloth square, 35 × 35 cm
- Shoe box, with lid cut in half
- Screening to cover beaker, shoe box
- Lamp
- Self-locking one-quart plastic freezer bags (2)
- Tap water, hot
- Masking tape

Focus Question

Design your own **Focus Question** based on the information presented in the Introduction and Materials sections above, as well as the additional information presented in the Procedure section that follows. Write your Focus Question on the Vee Form.

Knowing Side

With the help of your teacher, your class will discuss the **Subject Area** background for this investigation. Then, with the help of your classmates, list the **Concepts** and the new **Vocabulary** words on your Vee Form. In the **Concept Statements** section of your Vee Form, use these words in sentences that define and explain them.

Procedure

Live crickets can be purchased locally from pet or bait shops, or ordered from Ward's.
Crickets
(87 M 6100) pkg. of 10

Part A: External Structure

1. **CAUTION: You will be working with a live animal. Be sure to treat it gently and to follow directions carefully.**
2. To slow the activity of your cricket, place a beaker containing a cricket into an ice bucket or insulated cooler. Fill the bucket or cooler with enough crushed ice to completely surround the beaker. After about 15 minutes, your cricket should be inactive. Carefully remove the cricket from the beaker to begin examining its external structures. If your cricket becomes too active before you are finished examining it, put it back into the beaker in the ice bucket or cooler and wait until it is quiet and still again.
3. One characteristic of insects that sets them apart from other invertebrates is that their bodies are divided into three sections. Label the three body sections on the diagram of the cricket in the **Additional Records and Observations** section on the back of your Vee Form.

Caution students to examine their crickets carefully, to avoid damaging them. You may wish to provide students with a few cricket specimens in a sealed container that have been preserved in 70% ethyl alcohol. It will then be easier for students to examine body parts.

4. Determine the sex of your cricket by comparing it to the illustrations below of the male and female cricket.

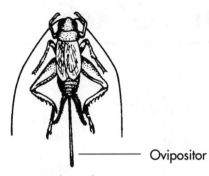

Ovipositor

Male cricket   Female cricket

Is your cricket a male or a female? Explain.

Answers will vary. Female crickets have a long ovipositor at the end of their abdomens.

Males do not.

5. Label the head on the cricket diagram. Use a hand lens or stereomicroscope to examine the cricket's head and mouthparts.

Describe the mouthparts of the cricket shown on the cricket diagram.

labrum: an upper lip, located anterior to the mandible

mandibles: short and relatively thick, for chewing
maxillae: longer and segmented, posterior to the mandibles

labium: segmented, a lower lip

6. Label the thorax on the cricket diagram.

Label both the walking legs and the jumping legs on the cricket diagram.

How do the cricket's jumping legs differ from its walking legs?

The femur on the jumping leg is much larger than the one on the walking leg.

Examine the tibia (bottom part of foreleg) of one of the forelegs with your hand lens. You should be able to see a tympanum located on the outer edge. Label the tympanum on the cricket diagram.

What is the function of the tympanum? If necessary, look up "tympanum" in a dictionary.

hearing

Compare the wings of the male and female cricket.

What differences do you observe?

The wings of the female are smaller and have finer veins than those of the male.

Use your hand lens or the stereomicroscope to closely examine the veins on the underside of the male's forewings, and compare them to the diagram below.

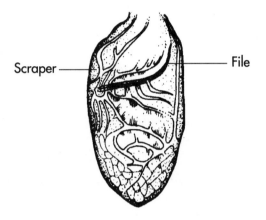

Scraper — — File

Forewing of male cricket

What role does the cricket's wing design play in producing the chirping sound made by male crickets?

Male crickets chirp by pulling a heavy vein on the wing with many sharp ridges or teeth (file) across the "scraper," which is a hardened vein at the edge of the wing.

7. Label the abdomen on the cricket diagram.

How many abdominal segments does your cricket have?

10

Examine the ovipositor at the posterior end of the abdomen of a female cricket. Find the cerci at the end of the abdomen. These structures are similar to antennae, in that they can sense movements in the air or on the ground. Label the cerci on the cricket diagram.

To which abdominal segment are the cerci attached?

the tenth

Part B: Behavior

1. Place a different cricket in a 500-mL beaker and cover the beaker with a piece of screening. As you perform the following procedure steps, record your observations in the table in the **Additional Records and Observations** section of your Vee Form.

2. To observe your cricket's behavior, sit very quietly as you watch your cricket. Any sudden movements will alarm the cricket and it will stop whatever it is doing. Slowly add a tiny piece of apple to the beaker.

How does the cricket respond?

It walks over to the apple and examines it.

Use a hand lens to determine how the cricket examines the fruit, holds it, and eats it.

What mouthparts are used to hold food?

labrum and maxillae

What mouthparts are used to chew food?

mandibles

3. An important means of communication among crickets is chirping. In most species, the ability to chirp is found only in males, who have four different songs. The calling song is the one most often heard. Males use it to attract females. After the attention of a female is captured, the male sings a courtship song. The male sings a rivalry song when another male invades its territory. It sings an alarm song when a predator is nearby.

   If your cricket is a female, find another student who has a male. If the male is not already chirping, cover the beaker with a cloth square and wait quietly. When the cricket begins chirping, slowly lift the cloth and observe it.

Describe the actions of the cricket when it chirps.

It holds its wings up and rubs them together. Male crickets chirp by rubbing their two forewings together.

Place the beaker containing the chirping male in a large shallow cardboard box or an empty 10-gallon aquarium. Release a female cricket in the end of the box or aquarium that is opposite from the beaker containing the chirping male.

How does the female respond to the chirping male?

Answers may vary, but often the female moves in a straight line toward the male.

Tap the beaker so that the male cricket stops chirping.

How does the female respond when the male stops chirping?

Answers may vary, but often the female wanders randomly.

4. To observe how crickets respond to light, place your cricket in a shoe box and cover the box with screening. Put the half lid over one end of the shoe box and place a light directly over the other end. Wait five minutes before looking inside the box.

On which side of the box is the cricket?

the dark side

Move the half lid and the light to the other end of the box, and wait five minutes.

How does your cricket respond to the light?

Most crickets will move away from the light.

5. Set up the apparatus described below to observe how a cricket responds to temperature. Place your cricket in a 500-mL beaker. Cover the beaker with another 500-mL beaker, so that the spouts of the two beakers are aligned. Join the two beakers together with masking tape. Fill one plastic bag halfway with crushed ice. Fill another plastic bag halfway with hot tap water. Seal each bag securely, and place the bags together as shown in the diagram. Gently rock the joined beakers from side to side until the cricket is in the area where the two beakers meet. Place the beakers on top of the plastic bags, using strips of tape to hold them to the bags, if necessary. Wait five minutes before checking your cricket's location. Repeat the experiment several times, placing the cricket in different areas of the beakers each time.

Describe the cricket's behavior.

The cricket will consistently move to the same side, usually the side with the warmer

temperature.

What temperature does your cricket seem to prefer?

warmer temperatures

6. Clean up your materials and wash your hands before leaving the lab.

Use the information on the **Knowing Side** of the Vee to interpret your results from the **Doing Side**, and then write your **Knowledge Claim.** Write a **Value Claim** for this study.

Analysis

When the lab is completed, students may release crickets captured in the wild to the areas where they were collected. Crickets purchased from pet shops or bait shops may be returned to the place of purchase or used as food for amphibians or reptiles in upcoming investigations. Crickets can be kept alive in the classroom for several weeks in an aquarium containing a layer of sawdust. You will need to provide a shallow water dish containing a sponge, occasional slices of apple or potato, and dry dog food.

1. What characteristics of the cricket cause it to be classified as an insect?

a body divided into three distinct parts and three pairs of jointed legs

2. Describe why crickets belong to the order Orthoptera.

Crickets have two pairs of straight wings and they have legs that are modified for jumping.

3. How does chirping help crickets survive?

It helps them find mates and may also alert other crickets to the presence of predators.

4. How might their ability to perceive temperatures help crickets to find shelter?

By sensing which shelter is warmer, they can move to one more suitable for them.

Going Further

In the appropriate space on your Vee Form, write a **New Focus Question** that could be the point of a new investigation based on what you have just learned.

• Male crickets will stake out a territory and defend it against any intruding male. To observe territorial behavior in male crickets, put a layer of sand across the bottom of an aquarium and scatter some twigs, rocks, and bark in different areas to provide hiding places. Put two male crickets in the aquarium and allow them to get settled and establish a territory. Introduce a third male one day later and observe how the crickets behave. Remove the third male and introduce a female cricket when the original males get resettled. Compare the behavior of the males when a female is introduced with their behavior when a male is introduced.

Name _____ Date _____ Class _____

Live Crickets

KNOWING SIDE

DOING SIDE

Focus Question:

What are the external structures and behaviors of crickets?

New Focus Question:

How do crickets compare to other similar insects, for example, a grasshopper?

Subject Area:

Classification
Ethology (animal behavior)

Concepts:

behavioral response
chirping calls
courtship

Vocabulary:

Orthoptera
ovipositor
labrum
mandible
maxilla
labium
tibia
tympanum
cerci

Concept Statements:

1. House and field crickets chirp any time of day or night and rarely fly.
2. Crickets are in the insect order Orthoptera, which means they have two sets of wings.
3. Crickets have jumping hindlimbs.
4. Female crickets and male crickets find each other to mate.
5. Male crickets have four calls, depending on the purpose: attracting a female, courting a female, a territorial call, and an alarm call in the presence of a predator.

Value Claim:

Crickets can be found in dark, warm places, such as under a rock or log.

Knowledge Claim:

Most crickets prefer darkness and warm temperatures. The female is attracted to the chirping male as it approaches the vicinity of the male.

Records:

[See Additional Records and Observations on the following page]

Materials:

live crickets, crushed ice, ice bucket, 500-mL beakers (3), apple, hand lens or stereomicroscope, 35 × 35 cm cloth square, shoebox with lid cut in half, screening to cover beaker and shoebox, lamp, self-locking 1-quart freezer bags (2), hot tap water, masking tape

Procedure:

Identify cricket's external structures according to drawings and observation. Study chirping in males and responses in females. Observe responses to light and temperature.

Additional Records and Observations:

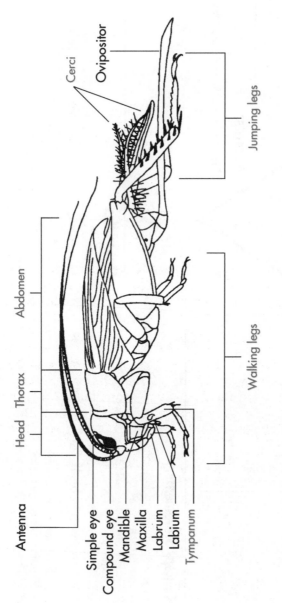

Antenna

Head   Thorax   Abdomen

Simple eye
Compound eye
Mandible
Maxilla
Labrum
Labium
Tympanum

Cerci

Ovipositor

Jumping legs

Walking legs

Cricket Observations

Behavior	Male	Female
Feeding		
Chirping		
Reaction to light		
Reaction to cold/heat		
Movement		
Other behavior		

Comparison of Five Arthropod Classes

Introduction

All live and preserved specimens may be purchased from Ward's. You may substitute other specimens for those specified in the investigation. If you wish, collect live specimens native to your area.
For Ward's catalog numbers for the suggested specimens, see the annotation following Going Further.

Materials

The phylum Arthropoda is the most successful and diverse of all the phyla in the animal kingdom. Although arthropods are related to the earthworm, evidenced by their segmented bodies, much of the success of the arthropods can be attributed to their more complex nervous systems, grouped muscles, jointed legs, wings, and hard exoskeletons.

In this investigation you will be able to see some of the more specialized parts of arthropod bodies that help them adapt to almost any environment. The five major characteristics that all arthropods have in common are jointed appendages, segmented bodies, exoskeletons, dorsal hearts and ventral nervous systems. Each of the five classes of arthropods has special characteristics that set it apart from the others. In studying these classes, you will observe the characteristics that all the arthropods have in common as well as how they differ.

- Live specimens:
  Crayfish
  Tarantula
  Cricket
  Millipede
  Pill bugs
- Boxes (2), clear plastic
- Water, dechlorinated
- Beaker, 500-mL
- Screening, 10 cm sq.
- Blunt probe
- Preserved specimens:
  Centipede
  Tarantula
- Watch glass
- Paper towels

Focus Question

Design your own **Focus Question** based on the information presented in the Introduction and Materials sections above, as well as the additional information presented in the Procedure section that follows. Write your Focus Question on the Vee Form.

Knowing Side

With the help of your teacher, your class will discuss the **Subject Area** background for this investigation. Then, with the help of your classmates, list the **Concepts** and the new **Vocabulary** words on your Vee Form. In the **Concept Statements** section of your Vee Form, use these words in sentences that define and explain them.

Procedure

1. **CAUTION: You will be working with live animals. Be sure to treat them gently and to follow all directions carefully**. Obtain a live crayfish and place it in a plastic box containing 2 to 3 centimeters of dechlorinated water. Examine your crayfish for the number of body sections, the number of legs, the presence of mandibles, and the type of respiration.

Record your observations in the table in the **Additional Records and Observations** section on the back of the Vee Form. Enter in the table the name of the class to which a crayfish belongs. You may use your textbook during this investigation to help you determine the presence or absence of mandibles and the type of respiratory organs.

Students should not handle the preserved specimens. You may want to place a single specimen on a watch glass under a dissecting scope, for viewing by the entire class.

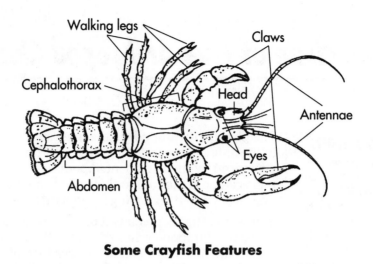

Some Crayfish Features

How does the type of respiration used by the crayfish reflect an adaptation to its environment?

Gills are an effective respiratory structure in water, the crayfish's habitat.

2. Observe the tarantula in the class terrarium. Do not attempt to remove it. Obtain a preserved tarantula in a sealed container and examine it carefully. Record your observations in the table.

What respiratory structure is found in arachnids that is not found in other arthropods?

book lungs

Where are they located?

anterior ventral surface of the abdomen

Compare your crayfish with the tarantula.

What features do the crayfish and tarantula share?

Both have jointed legs, an exoskeleton, a segmented body, two main body segments, and eight legs used for walking.

What features of the crayfish and tarantula are different?

The crayfish has gills, antennae, mandibles, and modified legs not used for walking (including chelipeds for protection and food gathering, and swimmerets for swimming). The tarantula has book lungs, tracheae, and malpighian tubules.

3. Obtain a live cricket from your teacher and place it in a 500-mL beaker. Cover the beaker with screening. Observe the cricket and record your observations in the table.

Compare and contrast the legs of the cricket with those of the crayfish.

Cricket leg modifications include enlarged, muscular hind legs for jumping. Crayfish legs are designed for walking on the bottoms of streams and lakes.

4. Line the bottom of the second plastic box with a damp paper towel. Obtain a live millipede from your teacher and place it in the box. Wait a few moments until the millipede relaxes and begins to move about. Examine the millipede and record your observations in the table on the back of the Vee.

How many legs does the millipede have on each segment?

two pairs of legs

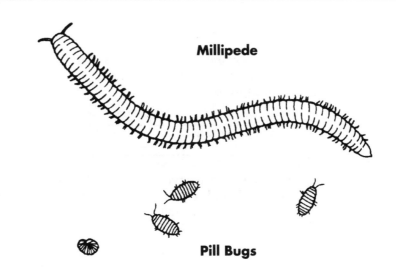

Millipede

Pill Bugs

Describe the way in which the millipede moves.

Its legs move with a wavelike motion.

5. Crustaceans are a highly diverse group of organisms. Pill bugs are unusual crustaceans because they adapted to living on land. Although pill bugs and millipedes belong to different classes of arthropods, they resemble each other in general appearance.

Obtain a live pill bug and place it in the box with the millipede. Allow the pill bug a few minutes to become accustomed to its new surroundings. Make sure the paper towel is not allowed to dry out. Prod the millipede and pill bug very gently with a blunt probe.

What is the behavioral response of each animal when it is disturbed?

They curl up. The pill bug makes a ball, the millipede a spiral or oval.

What characteristic shared by all arthropods makes this behavioral response an effective defense mechanism for these animals?

hard exoskeleton

What characteristic shared by pill bugs and millipedes enables them to exhibit this behavioral response?

Many small similar segments make these animals flexible enough to curl up.

What type of respiration would you expect the pill bug to have?

Pill bugs breathe by means of tracheae.

How does respiration in the pill bug differ from respiration in the crayfish? Offer an explanation for the different forms of respiration exhibited by these two organisms.

Pill bugs are terrestrial and thus respire by means of tracheae. Crayfish are aquatic and thus respire by means of gills.

6. Obtain a preserved centipede in a sealed container. Examine the centipede and record your observations in the chart.
7. Clean up your materials and wash your hands before leaving the lab.

Use the information on the **Knowing Side** of the Vee to interpret your results from the **Doing Side**, and then write your **Knowledge Claim**. Write a **Value Claim** for this study.

Analysis

1. For each of the characteristics below and on the following page, write "all arthropods" next to the characteristic if it is shared by all arthropods. If the characteristic is not shared by all arthropods, write the name of the class or classes that show that characteristic.

Jointed legs   all arthropods

Exoskeleton   all arthropods

Segmented body   all arthropods

Ventral nerve cord   all arthropods

Open circulatory system   all arthropods

Dorsal heart	all arthropods
Tracheae	some crustaceans, Arachnida, Insecta, Chilopoda
Mandibles	Diplopoda, Crustacea, Insecta, Chilopoda
Malpighian tubules	Arachnida, Insecta, Chilopoda, Diplopoda
Two main body sections	Crustacea, Insecta, Arachnida
Gills	most crustaceans
Book lungs	Arachnida
Six legs	Insecta
Swimmerets	some crustaceans

2. Name one advantage and two disadvantages of an exoskeleton.

Advantage: protection from predators; disadvantages: the animal has to molt because the exoskeleton does not grow and the exoskeleton is heavy and therefore limits the size and activities of the animal.

3. What characteristics of the spider make it well adapted to survive on land?

The spider absorbs oxygen by means of book lungs and tracheae. Oxygen enters the spider's body through slits in the exoskeleton called spiracles. Most spiders catch prey in sticky webs. Spiders have eight walking legs.

4. Which one of the five classes of arthropods is the most successful? Explain.

Insecta, because they are so diverse. They can fly, swim, sting, live in colonies, camouflage themselves, live almost anywhere, and eat a variety of foods.

Going Further

In the appropriate space on your Vee Form, write a **New Focus Question** that could be the point of a new investigation based on what you have just learned.

- Make an in-depth study comparing the external and internal anatomies of the crayfish and the cricket. Design a chart that will compare all the body parts. Write a B next to the parts common to both, a C next to parts belonging to only the crayfish, and an I next to parts belonging to only the cricket. You may also wish to color-code the body parts.

The following specimens may be obtained from Ward's.

Live:

Crayfish
(87 M 6031) pkg. of 12

Tarantula
(87 M 6965)

Crickets
(87 M 6100) pkg. of 10

Giant Tropical Millipede
(87 M 6970)

Isopods (Pill Bugs)
(87 M 5520)

Preserved:

Large Southern Centipede
(68 M 3112) jar of 10

Tarantula
(68 M 3071)

Name _____ Date _____ Class _____

Comparison of Five Arthropod Classes

KNOWING SIDE

DOING SIDE

Focus Question:

How do the five classes of arthropods differ and what do they have in common?

Value Claim:

Any newly discovered arthropod organism could be compared to these five classes for classification.

New Focus Question:

How do the internal characteristics of different arthropod classes compare?

Knowledge Claim:

All five classes of arthropods have jointed appendages, but differences appear in the number of body segments, number of legs, and method of respiration, for example.

Records:

[See Additional Records and Observations on the following page]

Subject Area:

Classification
Adaptation

Concepts:

characteristics
similarities
differences

Vocabulary:

arthropod
niche
appendages
exoskeleton

Materials:

Live specimens: crayfish, tarantula, cricket, millipede, pill bugs; preserved specimens: centipede, tarantula; clear plastic boxes (2), dechlorinated water, 500-mL beaker, 10 cm sq. screening, watch glass, paper towels

Procedure:

Observe live and preserved specimens of phylum Arthropoda, and compare and contrast them in terms of their different abaptations and habitats.

Concept Statements:

1. The arthropods' segmented bodies relate them to earthworms.
2. Arthropod means jointed appendages.
3. More advanced characteristics allow arthropods to respond more effectively to their environments.
4. All arthropods have jointed appendages, segmented bodies, exoskeletons, dorsal hearts and ventral nervous systems.

Additional Records and Observations:

Arthropod Comparisons

Class	Example	Body Sections	Number of Legs	Mandibles	Respiration
	crayfish				
	tarantula				
	cricket				
	millipede				
	centipede				

Fish Morphology and Behavior

Introduction

Fishes are vertebrates classified in the phylum Chordata. Unlike arthropods, fishes have a backbone and an internal skeleton. An internal skeleton does not restrict movement and makes it possible for animals to grow larger without the need to shed an outgrown skeleton. Though some invertebrates have body tissues forming organs, the organs in fishes are much more highly developed and specialized than those of invertebrates.

Most fishes are members of the class Osteichthyes, the bony fishes. In this investigation you will examine the external features and behavior of the goldfish, which is a bony fish. You will also observe the diversity among bony fishes by using a dichotomous key to classify members of a group of representative fishes.

Materials

- Beaker, 500 mL
- Water, dechlorinated
- Fish net, small
- Goldfish, live
- Box, cardboard, to fit over beaker
- Flashlight
- Largemouth bass or carp, fresh or preserved
- Forceps
- Microscope slide
- Medicine dropper
- Water, tap
- Stereomicroscope
- Prepared slides:
  Ctenoid scales
  Cycloid scales
  Ganoid scales
  Placoid scales

Focus Question

Design your own **Focus Question** based on the information presented in the Introduction and Materials sections above, as well as the additional information presented in the Procedure section that follows. Write your Focus Question on the Vee Form.

Knowing Side

With the help of your teacher, your class will discuss the **Subject Area** background for this investigation. Then, with the help of your classmates, list the **Concepts** and the new **Vocabulary** words on your Vee Form. In the **Concept Statements** section of your Vee Form, use these words in sentences that define and explain them.

Procedure

Part A: Morphology and Behavior

1. **CAUTION: You will be working with a live animal. Be sure to treat it gently and follow directions carefully.**
2. Fill a 500-mL beaker half full with dechlorinated water that is about the same temperature as that of the aquarium water. Wet the fish net first, then catch a goldfish from the aquarium by putting the net under the water, gradually cornering the fish, and then promptly lifting the net straight up. DO NOT press the fish against the side of the aquarium and slide it up. Invert the net over the beaker and gently release the fish. Make sure that there is enough water to cover the dorsal fin and to allow the fish to swim freely.
3. Closely examine your fish, paying close attention to its external features. Compare your goldfish with the labeled diagram of the fish below, which shows some common external features of bony fishes. Use the diagram to help you identify the external features of your goldfish.

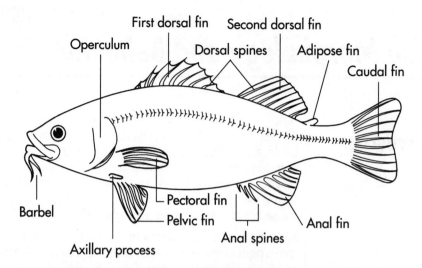

Operculum · First dorsal fin · Second dorsal fin · Dorsal spines · Adipose fin · Caudal fin · Barbel · Axillary process · Pectoral fin · Pelvic fin · Anal spines · Anal fin

Make a drawing of your goldfish in the **Records** section of your Vee Form, and label the following parts: mouth, eyes, gills, operculum, scales, pectoral fin, dorsal fin, anal fin, and caudal fin.

In what ways does your goldfish differ from the fish in the illustration above?

The goldfish has only one dorsal fin. It lacks barbels, dorsal spines, anal spines, and an axillary process. Its caudal fin is forked.

_____

_____

_____

4. Watch your fish as it swims. Pay special attention to the way the fish moves its fins and body.

Describe how the fish uses its fins and body to swim.

The fish moves its body back and forth as it swings its tail in the opposing direction. The paired fins help it change direction and the caudal fin pushes the fish forward.

_____

_____

_____

5. Look at the fish's eyes. Notice that it does not have eyelids. Fish do not need eyelids because their eyes are constantly moistened by water.

Describe exactly where the eyes are located on the fish.

on the sides of the head

_____

_____

_____

_____

6. Fish determine the vertical direction by responding to both gravity and light. A small inner ear organ, called the otolith, senses when the fish tilts from its vertical axis. Its eyes sense the direction from which light comes. Place the cardboard box with the open end facing up. Put the beaker into the box and tilt the beaker slightly to the side.

Describe the fish's orientation in the beaker. What did the fish do when you tilted the beaker?

The fish probably will remain in a vertical position.

Take the beaker out of the box. Turn the box so that the open end is on a side. Put the beaker into the box. The box will keep light from entering the beaker from the top. Shine the flashlight on the side of the beaker and observe the fish's initial reaction.

How does the fish first orient itself when the light is shone on the side of the beaker?

At first, the fish probably tilts slightly toward the light. Then the fish probably reorients itself in response to gravity so that it is vertical again.

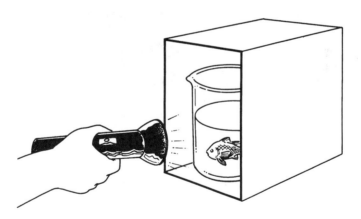

Record your observations of fish responses to gravity and light in the **Additional Records and Observations** section on the back of your Vee Form.

7. Return your goldfish to the aquarium by gently submerging the beaker and slowly tilting it until the fish is free.
8. From your teacher, obtain a large fish scale that has been removed from the middle of the body of a largemouth bass or carp. Clean the scale by rubbing it between your fingers in running water. Place the scale on a microscope slide and cover it with a few drops of water. Look at the scale, using low power on a stereomicroscope. Make a drawing of the scale in the space provided on the back of your Vee Form.

Fishes regenerate scales to replace those that are lost. Regenerated scales cannot be used to determine the age of a fish because they grew rapidly to fill in the vacant area.

9. Scales are formed by concentric layers of bone called circuli that are deposited at the edge of the scale as the fish grows. Circuli are closer together during periods of slow growth than during periods of fast growth. Among fishes of temperate regions, the resumption of fast growth each spring causes an annulus, which is a ring of circuli that grows around the scale. Determine the age of the fish by counting the number of annuli on the scale.

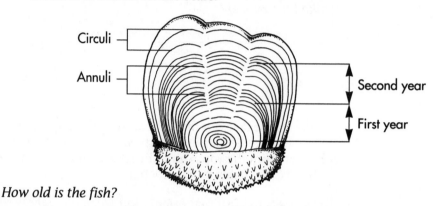

How old is the fish?

Answers will vary. Its age is equal to the number of annuli on the scale.

Large ctenoid scales can be obtained from large-mouth bass or carp. Scales from specimens that inhabit temperate regions are easiest to read. Trout have very small scales; a compound microscope will be necessary to count their annuli.

10. Use the stereomicroscope to look at the prepared slides of the four types of fish scales—ctenoid, cycloid, ganoid, and placoid. Compare the scale you examined with the scales on the prepared slides.

What kind of scales does your fish have?

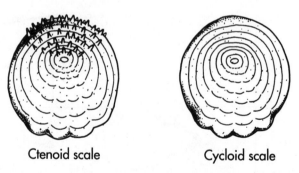

Ctenoid scale   Cycloid scale

Answers will vary. Ctenoid scales are usually found on spiny-rayed fishes such as bass, carp, sunfish, and perch. Cycloid scales are usually found on soft-rayed fishes, such as smelt, salmon, and trout. However, both ctenoid and cycloid scales may be found on the same fish.

Placoid scales are found exclusively on cartilaginous fishes, while ganoid scales are found only on gars and sturgeons.

Ctenoid scales and cycloid scales are the two types of scales that are most commonly found among bony fishes. Sometimes both ctenoid and cycloid scales can be found on the same fish.

11. Clean up your materials and wash your hands before leaving the lab.

Part B: Dichotomous Key

1. Carefully study the illustrations of the 10 different species of bony fishes on the next page.
2. Use the dichotomous key below to identify the 10 different species of fishes. Write their names in the spaces provided on the back of the Vee Form.

Use the information on the **Knowing Side** of the Vee to interpret your results from the **Doing Side**, and then write your **Knowledge Claim.** Write a **Value Claim** for this study.

Dichotomous Key to 10 Species of Bony Fishes	
1a Vertebral column does not extend into upper lobe of caudal fin	**2**
1b Vertebral column extends into upper lobe of caudal fin	**Lake sturgeon**
2a Adipose fin present	**3**
2b Adipose fin absent	**6**
3a Barbels absent	**4**
3b Barbels present	**Channel catfish**
4a Caudal fin forked	**5**
4b Caudal fin not forked	**Rainbow trout**
5a Axillary process at base of pelvic fin absent	**Rainbow smelt**
5b Axillary process at base of pelvic fin present	**Pink salmon**
6a Dorsal spines absent	**7**
6b Dorsal spines present	**8**
7a Dorsal fin long	**Starry flounder**
7b Dorsal fin short	**Northern pike**
8a Elongated body shape	**9**
8b Oval body shape	**Bluegill**
9a Three or more anal spines	**Largemouth bass**
9b Fewer than three anal spines	**Yellow perch**

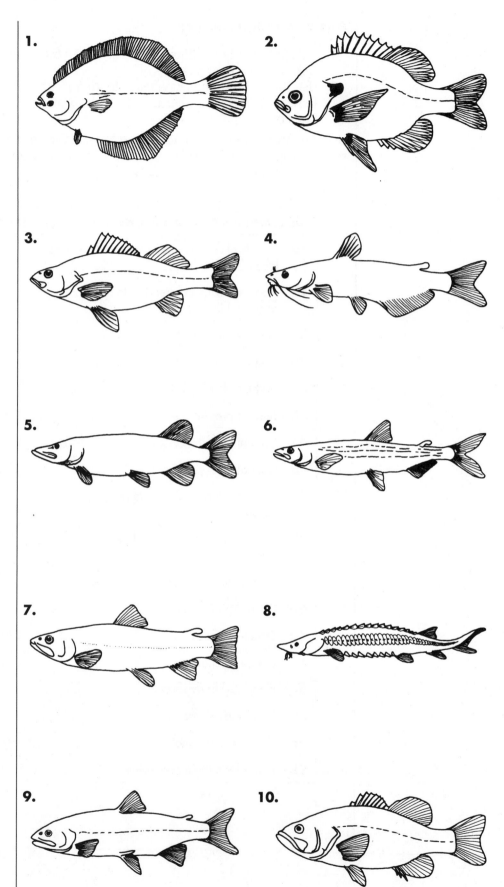

Analysis

1. Why is it advantageous for fishes to have their eyes located on the sides of their heads?

The eyes work independently, so that their field of vision covers both right and left as well as front and back at the same time. This is especially advantageous because fishes cannot turn their heads from side to side.

2. From your observations, do fishes seem to rely on gravity or on light to determine the upward direction?

Fishes rely primarily on gravity to determine which way is up. This was observed when the fish reoriented itself to a vertical position after briefly tilting toward the light.

3. Compare and contrast the three classes of fishes.

Jawless fishes have smooth, cylindrical bodies with skeletons formed of cartilage and sucker-like, jawless mouths. Cartilaginous fishes have bodies with skeletons formed of cartilage, hinged jaws with rows of teeth, and scaly skin. Bony fishes have skeletons made of bone, hinged jaws, and scaly skin.

4. What is the purpose of the swim bladder, and how does it work?

The swim bladder acts as a flotation device. As it fills with gases from the blood, the fish becomes more buoyant and rises in the water. As the swim bladder loses air, the fish becomes less buoyant and sinks.

5. What are some characteristics that separate different species of fishes?

Answers may include: shape of caudal fin, presence of adipose fin, presence of barbels, presence of axillary process at base of pelvic fin, presence of dorsal spines, and shape of body.

Going Further

In the appropriate space on your Vee Form, write a **New Focus Question** that could be the point of a new investigation based on what you have just learned.

- Condition a goldfish so that it will come to the surface of the water every time a bell is rung.
- Set up an aquarium and observe how the different kinds of fishes interact with one another. Notice where particular fishes usually swim in the aquarium and the positions of their mouths. Observe the behaviors of schooling fishes and solitary fishes.

Name _____ Date _____ Class _____

Fish Morphology and Behavior

KNOWING SIDE

DOING SIDE

Focus Question:

What are the external features and behaviors of fish? How can a dichotomous key be used to identify common North American fishes?

Subject Area:

Classification
Animal behavior

New Focus Question:

Can a goldfish be conditioned to come to the surface of the water every time a bell is rung?

Value Claim:

Being able to identify fish by external structures and behaviors can be fun at an aquarium.

Knowledge Claim:

A goldfish has one dorsal fin and a forked caudal fin. It moves by swinging its tail in the opposite direction of its body movement. Fish rely primarily on gravity (and to some extent light) to orient themselves in the water. The age of a fish can be determined by counting the number of annuli on its scales. Different types of fishes have different scale types.

Concepts:

morphology
internal skeleton

Vocabulary:

Chordata
Osteichthyes
vertebrate
circuli
annuli

Records:

Goldfish Sketch:

Materials:

500-mL beaker, dechlorinated water, fish net, live goldfish, cardboard box, flashlight, largemouth bass or carp, forceps, microscope slide, medicine dropper, stereomicroscope, prepared slides of cycloid, ctenoid, ganoid, and placoid scales

Concept Statements:

1. Most fishes are bony fishes and are in the class Osteichthyes and the phylum Chordata.
2. An internal skeleton does not restrict movement, and makes it possible for the fish to grow larger without molting.
3. Bony fishes, as a group, are very diverse.
4. Circuli are concentric layers of bone that form scales.
5. Behavior can be observed and described

Procedure:

A. Examine and identify external goldfish features and draw and label the observations. Describe its swimming behaviors and how it orients itself with respect to gravity and light. Compare different types of fish scales. B. Identify 10 different fish, using a dichotomous key.

Part A

Fish Responses to Gravity and Light

Condition	Response
Gravity	
Light	

Fish Scale Drawing:

Fish Scale Comparisons

Scale Type	Found on:	Descriptions
Ctenoid	Spiny-rayed fishes	
Cycloid	Soft-rayed fishes	
Placoid	Cartilaginous fishes	
Ganoid	Gars and sturgeon	

Part B

Dichotomous Key Results

1. starry flounder
2. bluegill
3. yellow perch
4. channel catfish
5. northern pike

6. rainbow smelt
7. rainbow trout
8. lake sturgeon
9. pink salmon
10. largemouth bass

HRW material copyrighted under notice appearing earlier in this work.

Additional Records and Observations:

Investigation **26-1**

Field Identification of Flying Birds

Introduction

A suggested field guide is Birds of North America: A Guide to Field Identification, by Robbins, Bruun, Zim and Singer (Golden Press, N.Y.) 1983.

Materials

Focus Question

Knowing Side

Procedure

The ability to recognize birds in the field is a useful skill to acquire. Birds can be identified by their size, color, habitat, plumage, and songs. However, one of the simplest criteria for identifying birds is their outline when they are flying. This is called their silhouette. The bird's outline gives many clues to its identity. In this investigation, you will use a description chart to become familiar with the structural characteristics of birds, and a group of unknown silhouettes that you will identify using provided charts.

• Pen or pencil
• Optional: Field guide to birds

Design your own **Focus Question** based on the information presented in the Introduction and Materials sections above, as well as the additional information presented in the Procedure section that follows. Write your Focus Question on the Vee Form.

With the help of your teacher, your class will discuss the **Subject Area** background for this investigation. Then, with the help of your classmates, list the **Concepts** and the new **Vocabulary** words on your Vee Form. In the **Concept Statements** section of your Vee Form, use these words in sentences that define and explain them.

1. Often, when you look up at flying birds, the only thing that can be seen is their outline. A bird's outline can tell you a lot about the bird, if you know what to look for. Some basic characteristics used to identify birds in flight can be found in the table that follows on this and the next page. Read and study this information, so that you can use the descriptions to help identify the silhouettes later on in this investigation.

Wing Type	Explanation	Illustration
Pointed	The outermost feathers are longest, as in the gull.	
Rounded	The middle primaries are longest and the remaining ones are graduated, as in the hawk.	
Narrow	corresponds to pointed	
Broad	corresponds to rounded	
Slotted	Every other feather stands out, causing the wings to look like ridges, as in the hawk.	

Neck Length	Explanation	Illustration
Long	as long or longer than the body, as in the flamingo	
Short	shorter than the body	

Tail Length	Explanation	Illustration
Long	longer than the body	
Short	the same length or shorter than the body	

Tail Shape	Explanation	Illustration
Square	All the feathers are the same length, as in the hawk.	
Rounded	Feathers are successively shorter toward the tail, as in the crow.	
Pointed	The middle feathers are much longer than the others, as in the grackle.	
Graduated	similar to rounded, but the gradations are more abrupt, as in the cuckoo	
Emarginated	The feathers increase in length from the middle to the outermost with slight gradations, as in the finch.	
Forked	similar to emarginated, but with abrupt gradations, as in the tern	

Feet in Flight	Explanation	Illustration
Extended	extended posteriorly beyond the body in flight, as in the heron	
Retracted	drawn under the body during flight, as in the tern	

Which wing type characteristic would more likely occur on a predatory bird that does a lot of maneuvering in flight?

slotted

Which neck length would typically occur on a bird that catches fish from the shoreline?

long

Which bird's feet are probably held out behind the body in flight: a perching songbird or a long-legged crane or egret?

crane or egret

2. Using the chart below, match the characteristics of each of the 14 birds to the unknown silhouette outlines given in the **Additional Records and Observations** section on the back of your Vee Form. Write the name of the bird on the line provided next to its outline.

Field Identification Chart

Names of Birds	Wing Types						Neck Length		Tail Length		Tail Shapes						Feet in Flight	
	pointed	rounded	narrow	broad	slotted	unslotted	long	short	long	short	square	rounded	pointed	graduated	emarginated	forked	extended	retracted
Whistling swan	X		X			X	X			X				X				X
Cuckoo	X		X	X				X	X					X			X	
Bald eagle		X	X	X				X		X	X							X
Loon	X		X	X	X		X			X		X					X	
Swift	X		X	X	X			X		X		X						X
Grackle		X	X	X				X	X				X					X
Finch	X		X	X	X			X		X					X			X
Osprey	X		X	X	X			X		X		X						X
Tern	X		X	X	X			X	X							X		X
Coot	X		X	X			X			X		X					X	
Sandhill crane		X	X	X			X			X		X					X	
Screech owl		X	X	X				X		X	X							X
Cormorant		X	X	X			X			X			X					X
Great blue heron		X	X	X			X			X		X					X	

Use the information on the **Knowing Side** of the Vee to interpret your results from the **Doing Side**, and then write your **Knowledge Claim.** Write a **Value Claim** for this study.

Analysis

1. What other structures of a bird's anatomy could be studied for identification?

bill (beak), feet, nails, plumage

2. What is shown by number 11 on the back of the Vee Form?

It is a bat, which is not a bird, but rather a mammal.

3. Do you see a pattern of wing shape correlating to slotted and unslotted wings? Explain.

Yes; broad wings are usually slotted and rounded. Narrow wings are usually unslotted and pointed.

4. Why would this type of identification be a good one for beginning bird watchers to learn?

One often sees birds flying. Identification can be made without tools, preparations, or capturing them.

Going Further

In the appropriate space on your Vee Form, write a **New Focus Question** that could be the point of a new investigation based on what you have just learned.

- Taking along a field guide, go outside and watch birds. You can do this in many different environments: by the shore, in the woods, on the fence posts and wires, at the seacoast, in marshlands, along railroad tracks, or near any body of water. Identify as many birds as possible.
- Add more birds to your Field Identification Chart and search for unusual birds in your area.

Field Identification of Flying Birds

KNOWING SIDE

Focus Question:

From a description of structural characteristics of birds and a field identification chart, can silhouette drawings of birds in flight be correctly identified?

Subject Area:

Ornithology
Classification

New Focus Question:

Can a novice birdwatcher successfully use a field guide to identify unusual birds after doing this investigation?

Concepts:

field identification

Vocabulary:

silhouette
plumage

Concept Statements:

1. Birds can be identified by their size, color, habitat, plumage, and songs, and especially by their outlines when they are flying (silhouettes).
2. Birds are classified according to their similarities.

DOING SIDE

Value Claim:

Learning what birds to expect in certain habitats and knowing their characteristics can be valuable in identifying them in the wild.

Knowledge Claim:

Patterns of wings, necks, tail size and shape, and feet position in flight are used to easily identify birds from silhouettes.

Records:

[See Additional Records and Observations on the following page]

Materials:

pencil, field guide to birds (optional)

Procedure:

Read and interpret the table of structural characteristics of birds, and identify silhouettes of birds using a field identification chart.

1. _____ Swan _____

2. _____ Osprey _____

3. _____ Cormorant _____

4. _____ Grackle _____

5. _____ Owl _____

6. _____ Swift _____

7. _____ Tern _____

8. _____ Bald eagle _____

9. _____ Crane _____

10. _____ Loon _____

11. _____ Bat _____

12. _____ Finch _____

13. _____ Coot _____

14. _____ Cuckoo _____

15. _____ Great blue heron _____

Sweat Gland Activity

Introduction

Sweat glands cover the entire human body, except for the genitals and the lips. In the skin alone, there are about 3 million sweat glands. On the palms of your hands, there are nearly 3000 sweat glands per square inch.

In this investigation, you will be able to show the presence of sweat glands in the skin and how they are distributed. Then you can observe what happens to the activity of the sweat glands when the environmental temperature changes, and when body temperature changes.

Materials

You may obtain iodine soap solution from Ward's. **Iodine Prep Solution** (15 M 9829)

- Millimeter ruler
- Heat lamp
- Erasable bond paper (2-in. squares)
- Cotton swabs
- Alcohol solution (70–95%)
- Antibacterial soap (with iodine)
- Towel or cloth

Focus Question

Design your own **Focus Question** based on the information presented in the Introduction and Materials sections above, as well as the additional information presented in the Procedure section that follows. Write your Focus Question on the Vee Form.

Knowing Side

With the help of your teacher, your class will discuss the **Subject Area** background for this investigation. Then, with the help of your classmates, list the **Concepts** and the new **Vocabulary** words on your Vee Form. In the **Concept Statements** section of your Vee Form, use these words in sentences that define and explain them.

Part A: The Effect of Heat on Sweat Gland Activity

Procedure

Because of the nature of erasable bond paper (abrasiveness and starch content), it works best for this investigation.

Prepare or obtain a 1% iodine solution, which is safe for skin use. However, caution students not to get any in their eyes. If they do, flush with water and consult a doctor. Remind students that iodine will stain clothes.

The reason for using the alcohol solution is to cleanse away skin oils. A 95% alcohol solution is preferable, but a 70% one will be acceptable.

1. **CAUTION: Both the rubbing alcohol and iodine solution are eye irritants. Wear goggles in Part A and Part B.**
2. Each lab team should select one person to be the experimental subject. Choose an area on the skin of your subject where sweating occurs readily, such as the place just below the hairline on the forehead. Then pick another place where it doesn't usually occur, such as the inner side of the forearm. Disinfect an area of about 3 square inches on each place with the alcohol solution.
3. Using a cotton swab, paint a 2-in. by 2-in. patch on each of these areas with the iodine soap and let it dry completely. Then gently press a 2-in. square piece of the bond paper on each area for 30 seconds.
4. When you remove the paper, look closely for dots with a bluish tinge, which are caused by the reaction of the iodine moistened by an active sweat gland. The iodine reaction with the starch in the paper causes the blue color. Therefore, the dots show the location of the active sweat glands. Mark a 1-cm square where the dots seem to be most concentrated. Count the number of dots there and record this number in Table A in the **Additional Records and Observations** section on the back of your Vee Form. Record your counts in the "Before Heat Application" column.

5. Record in Table A the skin flush (normal or darker) of the subject's skin at the two testing sites.

6. **CAUTION: The heat lamp is very hot. Do not touch it or stand closer than three feet from it.**

7. Focus the heat lamp on the back of the subject's head and shoulders, but cover the neck and head with a towel or cloth so that the heat will not directly be focused on bare skin.

8. After the heat lamp is turned on, the subject should tell you when the exact moment of heat sensation occurs. Continue for 10 minutes, until you see a definite change in sweat gland activity. Using a fresh piece of bond paper, take another set of readings, as in steps 3, 4 and 5. Record your results in Table A on the Vee Form.

Part B: The Effect of Exercise on Sweat Gland Activity

1. Repeat the steps done in Part A, but now have the subject run up and down stairs or run in place for 5 minutes. Record these readings in Table B on the back of your Vee Form.

2. Clean up your materials and wash your hands and test areas thoroughly before leaving the lab.

Use the information on the **Knowing Side** of the Vee to interpret your results from the **Doing Side**, and then write your **Knowledge Claim.** Write a **Value Claim** for this study.

Analysis

1. Did any difference in skin flush occur when heat was applied? Explain.

With heat, the sweat glands become more active on the forehead, but not on the arm.

When gland activity increases, the skin seems to get darker where the iodine was placed.

2. What is the function of sweating, flushing of the skin, and blanching of the skin? How do these adaptations relate to homeostasis?

We sweat to regulate our body temperature. We blanch to route more circulation to the

inner organs, we flush to rid ourselves of heat by dilation of blood vessels near the skin surface. All these things happen to keep the body in a balanced state (homeostasis).

Going Further

In the appropriate space on your Vee Form, write a **New Focus Question** that could be the point of a new investigation based on what you have just learned.

Infrared heat lamps should never be used. Also, as the alcohol and iodine solutions are flammable, be sure their containers are kept at a safe distance from the lamp.

Sweat Gland Activity

KNOWING SIDE

DOING SIDE

Focus Question:

How can the presence and distribution of sweat glands in the skin be demonstrated? How can a change in the activity of the sweat glands be determined when there is a change in the environmental temperature or body temperature?

Subject Area:

Homeostasis
Physiology

Concepts:

negative feedback

Vocabulary:

thermoregulation
hypothalamus

New Focus Question:

What would happen to sweat gland activity if an ice pack was placed on the back of a person's neck?

Concept Statements:

1. The brain's hypothalamus controls body temperature regulation.
2. Skin nerve endings sensitive to temperature are the sensors for temperature regulation.
3. The body strives to remain in a state of balance (homeostasis), by negative feedback.
4. Physiology is the study of the body's functions.

Value Claim:

Knowing the functions of sweating enables us to better understand how the body maintains internal stability.

Knowledge Claim:

With iodine and a starch-based paper, the number of sweat glands on a small area of skin can be estimated. Environmental heat causes an increase in sweat gland activity. Body temperature heat produced during exercise also causes increased sweating.

Records:

[See Additional Records and Observations on the following page]

Materials:

heat lamp, 70% or 95% alcohol solution, 1% aqueous iodine, cotton swabs, 2-in. squares of erasable bond paper, millimeter ruler, towel or cloth

Procedure:

Select a team member as the subject. On both the forehead and inner forearm, clean a 2-in. square area with alcohol, to clear skin oils. Paint iodine on both and let it dry. Press bond paper on lightly. Remove and look for blue dots. Count the number of dots in 1-cm^2 area and record. Introduce a heat source, and exercise, and take the same kind of readings.

Additional Records and Observations:

Table A The Effect of Heat on Sweat Gland Activity

Subject: ____		Before Heat Application	After Heat Application
Skin flush	forehead	normal	darker
	forearm	normal	normal
Sweat gland activity # dots/cm²	forehead	few	many
	forearm	none	none

Table B The Effect of Exercise on Sweat Gland Activity

Subject: ____		Before Exercise	After Exercise
Skin flush	forehead	normal	darker
	forearm	normal	no change
Sweat gland activity # dots/cm²	forehead	few to none	many
	forearm	none	none

Taste and Smell Sense Perception

Introduction

Smell and taste are separate senses, yet they combine to provide the numerous flavors humans experience. The taste receptors are located on the tongue, and are sensitive to stimuli that produce bitter, salty, sour, and sweet tastes. The olfactory receptors of the nose are also sensitive to a broad range of stimuli. However, before humans can taste or smell in response to any stimulus, the stimulus must be dissolved in liquid. When you are tasting, the food stimulus is dissolved in saliva. When you are smelling, the food stimulus is dissolved in the mucus inside the nasal passages. In this investigation, you will become more familiar with the receptors for taste and smell, and the relationship between the two senses.

Materials

Part A
- Masking tape
- Pen
- Paper cups, small (7)
- Spoon
- Sucrose solution, 5%
- Sodium chloride solution, 10%
- Acetic acid solution, 1%
- Quinine sulfate solution, 0.1%
- Cotton swabs
- Graduated cylinder, 25 mL

Part B
- Stirring rod
- Potato homogenate
- Onion homogenate
- Apple homogenate

Focus Question

Design your own **Focus Question** based on the information presented in the Introduction and Materials sections above, as well as the additional information presented in the Procedure section that follows. Write your Focus Question on the Vee Form.

Knowing Side

With the help of your teacher, your class will discuss the **Subject Area** background for this investigation. Then, with the help of your classmates, list the **Concepts** and the new **Vocabulary** words on your Vee Form. In the **Concept Statements** section of your Vee Form, use these words in sentences that define and explain them.

Procedure

The sucrose, sodium chloride, and acetic acid solutions can be readily prepared. The quinine sulfate solution is available from Ward's.

Quinine Sulfate (37 M 9545) 0.1% solution

Students should be cautioned not to ingest any test materials.

Part A: Taste

1. Cut four pieces of masking tape for use as labels. Place one label on each of four paper cups and label them "sweet," "salty," "sour," and "bitter."
2. Fill the cups one-quarter full with their appropriate solutions:
   sweet—5% sucrose solution
   sour—1% acetic acid solution
   salty— 10% sodium chloride solution
   bitter—0.1% quinine sulfate solution
3. Choose one person to be the tester and one to be the taster. Have the taster close his/her eyes. The tester should dip a cotton swab in one of the solutions. Do not tell the taster which solution has been chosen.
4. Place the swab in turn on the sides, the tip, the front surface, and the back surface of the tongue of the taster, as shown on the tongue diagram on the following page.

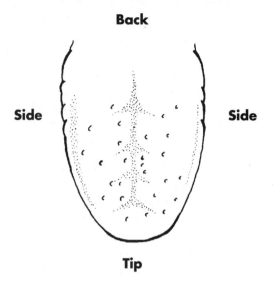

Back

Side

Side

Tip

5. Record responses of taste sensations in Table A of the **Additional Records and Observations** section on the back of your Vee Form. Use the following code:

   *** = very strong
   ** = moderate
   * = slight
   0 = no taste

6. Repeat the procedure to test the remaining solutions. Use a new cotton swab for each solution and discard the applicator, once the chemical has been applied to the tongue. Record all taste sensations in Table A on the back of your Vee Form.

7. Repeat the procedure, switching roles with your lab partner.

8. Place the remaining contents of the cup labeled "sweet" in a graduated cylinder. Pour half into a cup for you and half into a cup for your partner. Add the same amount of water to each cup, adding a volume equal to the sucrose solution volume in the cup. It has now been reduced to a 2.5% solution. Test to see if you can still distinguish the taste.

9. If you can still sense the sweetness after one dilution, dilute it by half again and repeat the test. This will be half of the previous dilution, or 1.25%.

10. Repeat the dilution step until you can no longer identify the sweet taste. Be sure to rinse your mouth out with water between dilutions!

11. Your threshold level concentration for a particular taste is between the concentration you cannot taste and the next higher concentration. Record the number of dilutions you had to make and your threshold level in Table B in the **Additional Records and Observations** section on the back of your Vee Form.

12. Repeat the dilution procedure for the three remaining solutions and record your results for each solution in Table B on the back of your Vee Form.

The homogenates can be prepared as follows:
a. Peel away and discard the respective outer surfaces.
b. Cut into small pieces
c. Place pieces in blender with one cup of water
d. Blend at high speed for about 5 minutes
e. Let stand for 5 minutes
f. Slowly pour off the clear liquid on top
g. Pour the homogenate that remains into a container.

Part B: Smell and Its Relation to Taste

1. Label three new cups "apple," "onion," and "potato." Fill each cup one-quarter full with the appropriate homogenate.
2. Select and blindfold a person and have him or her smell the homogenates. Bring the cups up to the nose of the tested person to determine if the substances can be identified correctly by smell.
3. Record the findings in Table C on the back of your Vee Form. Use a "+" if the response was correct and a "0" if incorrect.
4. Use a spoon to randomly feed a small amount of each homogenate to your lab partner, who should have his or her nose held shut, so as not to be able to smell the samples. Rinse the spoon after each trial, and have the taster rinse his/her mouth. Do not tell the taster the result of each test until all of Table C responses are recorded on his/her Vee Form.
5. Repeat the test, only this time let the taster's nose remain open. Record these findings under "Taste" in Table C on the back of your Vee Form.
6. Repeat the procedure, switching roles of taster/tester with your lab partner.
7. Clean up your materials and wash your hands before leaving the lab.

Use the information on the **Knowing Side** of the Vee to interpret your results from the **Doing Side**, and then write your **Knowledge Claim**. Write a **Value Claim** for this study.

Analysis

1. What do the results indicate about the senses of smell and taste? Explain.

   Taste and smell are closely associated. When food is tasted, it is smelled as well. This is evident when you have a head cold and various foods taste nearly the same, because you cannot smell very well.

2. In the **Records** section of your Vee Form, draw a diagram of the tongue and indicate where the receptors for each of the four tastes are located.
3. While there are only four basic tastes, many flavors are experienced. How is this possible?

   The tastes of the many flavors that can be identified result from a mixture of the four tastes and the combined effect of smell and taste together.

Going Further

In the appropriate space on your Vee Form, write a **New Focus Question** that could be the point of a new investigation based on what you have just learned.

• Determine the threshold levels of other common liquids, such as orange juice and milk.

Name _____ Date _____ Class _____

Taste and Smell Sense Perception

KNOWING SIDE

Focus Question:
What are the specific areas on the tongue for taste perception? How do the sensory perceptions of taste and smell relate to each other?

Subject Area:
Physiology
Sense perception

Concepts:
stimulus/response
threshold

Vocabulary:
olfactory
mucus

Concept Statements:
1. The threshold level for any sensation is between the concentration you cannot sense and the next higher concentration.
2. Olfactory receptors of the nose are sensitive to a broad range of stimuli.
3. When tasting, the stimulus is dissolved in saliva. When smelling, the stimulus is dissolved in the nasal mucus.

DOING SIDE

Value Claim:
In marketing, the food industry would benefit from knowing the average threshold level of the four basic tastes.

Knowledge Claim:
Taste receptors identify sweet at the tip of the tongue, salt at the front, sour at the sides, and bitter at the back. Flavors are a combination of these four tastes, and derive from a combination of smelling and tasting together.

Records:
Tongue Taste Receptors:

New Focus Question:
What might be the threshold levels of other common liquid foods, such as orange juice and milk?

Materials:
A. masking tape, pen, paper cups, spoon 5% sucrose solution, 10% NaCl solution, 1% acetic acid solution, 0.1% quinine sulfate solution, cotton swabs, 25-mL graduated cylinder
B. stirring rod; potato, onion, and apple homogenates

Procedure:
A. Discriminate among the four taste sensations and discover your threshold for each.
B. Differentiate between different food flavors by taste, smell, and by taste without smell.

Table A   Location of Taste Receptors

Taste	Area of Tongue			
	Sides	Tip	Front	Back
Sweet		*		
Sour	*			
Salty			*	
Bitter				*

Table B   Taste Thresholds

Solution	Number of Dilutions	Threshold Level
5% Sucrose		
10% Sodium chloride		
1% Acetic acid		
0.1% Quinine sulfate		

Table C   Relation of Taste to Smell

Homogenate	Smell	Taste	Taste Without Smell
Onion			
Potato			
Apple			

Effects of Hormones on Circulation

Introduction

Adrenaline and *Daphnia* can be purchased from Ward's.

Daphnia Magna (87 M 5210) Culture

Adrenaline Solution (38 M 2053) 1:10,000 (Aq.) The 1:100,000 solution (0.00001%) may be made from this by appropriate dilution.

The endocrine system is vital to homeostasis. It plays an important role in regulating the cellular activities in the body. Hormones are chemical messengers from the endocrine system organs that travel in the blood and cause changes that lead to growth and development, reproductive ability, and a balanced state of the many body systems. Each hormone is specific to a certain function and usually affects only one organ or group of organs. In order to cause a physiological change, the hormone has to be present in a sufficient amount. This is called the threshold level. If the amount present is less than threshold, the target organ will not show any change.

In this investigation, you will see the effects of adrenaline in different concentrations on the heartbeat rate of *Daphnia*. Commonly called "water flea," it is not really an insect, but rather a small crustacean. Due to its transparent body wall and pulsating heart, it is easy to observe increases or decreases in its heartbeat rate under the microscope. Since adrenaline affects the heartbeat rates of *Daphnia* and humans in a similar way, *Daphnia* make a good choice for the study of the effects of hormones on circulation.

Materials

Per team of two:
- *Daphnia culture*
- Petri dish
- Depression slide
- Stopwatch
- Paper towel
- Compound light microscope
- Eyedropper (wide mouthed)
- Plastic spoon
- Adrenaline solutions:
  0.0001% (0.1 mL)
  0.00001% (1 mL)

Focus Question

Design your own **Focus Question** based on the information presented in the Introduction and Materials sections above, as well as the additional information presented in the Procedure section that follows. Write your Focus Question on the Vee Form.

Knowing Side

With the help of your teacher, your class will discuss the **Subject Area** background for this investigation. Then, with the help of your classmates, list the **Concepts** and the new **Vocabulary** words on your Vee Form. In the **Concept Statements** section of your Vee Form, use these words in sentences that define and explain them.

Procedure

The heartbeat rate will be high during a 10 second interval, somewhere near 40 beats.

1. Gently remove several *Daphnia* from the class stock collection and place them in your petri dish. Transfer one of the *Daphnia* to a clean dry depression slide with your dropper, removing some of the excess water with a paper towel to contain the *Daphnia* in a small field of vision.

Daphnia

2. View the slide under the microscope at low power to find the *Daphnia*. Count its heartbeats while your partner counts the 10 seconds. The counter should locate the heart and tell the timer when to begin timing.

3. Record the number of heartbeats in Table A in the **Additional Records and Observations** section of your Vee Form. Repeat step 2 twice more and record your data in Table A.

4. Calculate the average heartbeat number per minute for your *Daphnia*, before any hormones have been added. Multiply each 10-second-count by six to get the number of beats per minute (BPM). Add all three BPM numbers and divide by three to obtain the average BPM.

5. Add a drop of the 0.00001% adrenaline solution to the *Daphnia* under your scope. Wait 30 seconds for the adrenaline to take effect. Repeat steps 2–4 above to determine the average BPM for that concentration of adrenaline, and record it in Table B on the back of your Vee Form.

6. Remove the slide and get a fresh *Daphnia*, then repeat the procedure with the next solution, which has 0.0001% adrenaline. Record your data in Table C in the **Additional Records** section. First be sure to get a control reading, that is, the heartbeat rate before adding any hormone.

7. Plot your data for each control reading and each adrenaline concentration for the three *Daphnia* in a graph, on the back of your Vee Form.

8. Clean up your materials and wash your hands before leaving the lab.

Place the different concentration adrenaline solutions at different areas of the room, so students will not mix them. Provide a place for used *Daphnia*, so that none become reused during the investigation.

Use the information on the **Knowing Side** of the Vee to interpret your results from the **Doing Side**, and then write your **Knowledge Claim**. Write a **Value Claim** for this study.

Analysis

1. Where is adrenaline produced in humans? Under what conditions?

   adrenal glands; stress

2. Which solution(s) of adrenaline caused an effect at or above a threshold level?

   the 0.0001% solution

3. If you administered an adrenaline solution with a concentration 100 times the threshold level, what do you think would happen to *Daphnia*?

   They would die.

Going Further

In the appropriate space on your Vee Form, write a **New Focus Question** that could be the point of a new investigation based on what you have just learned.

Name _____ Date _____ Class _____

Effects of Hormones on Circulation

KNOWING SIDE

DOING SIDE

Focus Question:

What are the effects of different concentrations of adrenaline on the heartbeat rate of *Daphnia* ?

Value Claim:

Understanding threshold levels for effects of hormones is crucial in areas of medicine such as the study of infertility.

Knowledge Claim:

The 0.0001% solution caused an increase in heartbeat rate, which speeds up the *Daphnia's* metabolism. The 0.00001% solution did not, and is thus below the threshold concentration.

Records:

[See Additional Records and Observations on the following page]

New Focus Question:

Can different concentrations of LH, which causes ovulation in mammals, cause certain cats to increase the number of kittens in their litters?

Subject Area:

Endocrinology
Homeostasis

Concepts:

hormones
threshold

Vocabulary:

adrenaline

Materials:

Per team of 2: 3 *Daphnia*, clock with second hand, paper towels, petri dish, slide, microscope, eyedropper (wide mouthed), 2 concentrations of adrenaline (0.0001%, 0.00001%)

Procedure:

Observe *Daphnia's* normal heart rate and compare to its heart rate at 2 differing concentrations of adrenaline. Graph the results.

Concept Statements:

1. Hormones are carried from endocrine glands to the body organs by blood and other body fluids, with no ducts.
2. Endocrine glands have target organs for hormone action.
3. Hormones bring about a physiological change in a target organ.
4. Threshold is the lowest level at which the hormone first has an effect.

Additional Records and Observations:

Table A Normal

Trial	# beats per 10 sec. × 6 = BPM
1	
2	
3	
Average	Avg. BPM

Table B Adrenaline 0.00001%

Trial	Beats/minute
1	
2	
3	
Average	Avg. BPM

Table C Adrenaline 0.0001%

Trial	Beats/minute
1	
2	
3	
Average	Avg. BPM

Adrenaline Effect on *Daphnia* Heart Rate

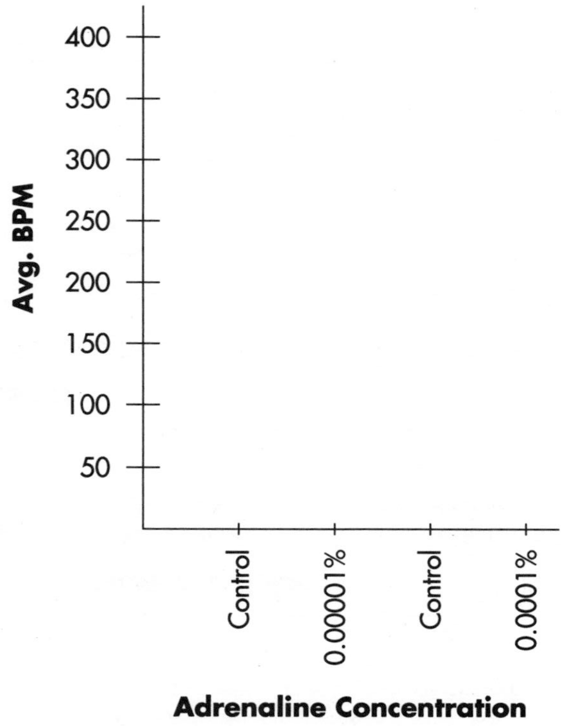

Avg. BPM: 50, 100, 150, 200, 250, 300, 350, 400

Adrenaline Concentration: Control, 0.00001%, Control, 0.0001%

How Drugs Affect the Heartbeat Rate

Introduction

Some drugs are stimulants that quicken the body's metabolism, while others are depressants that slow down this metabolism, for which heartbeat rate is an indicator. In this investigation, you will study the effects of several drugs on metabolism. The way the heartbeat rate responds to chemicals in *Daphnia* is similar to the response of the human heart to these same chemicals.

Materials

If the drug is not in liquid form, dilute it in water several hours before the investigation. *Daphnia* can be ordered from Ward's. See Investigation 29-1.

- Medicine droppers (10)
- *Daphnia*
- Depression slide
- Coverslip
- Microscope, compound light
- Paper towels
- Nicotine
- Watch (with second hand)
- Beaker, 100 mL
- Acetylcholine
- Aspirin
- Antacid
- Cough drops
- Coffee, freshly perked
- Nasal spray
- Ethyl alcohol, 10.0%
- Tea, freshly brewed

Focus Question

Design your own **Focus Question** based on the information presented in the Introduction and Materials sections above, as well as the additional information presented in the Procedure section that follows. Write your Focus Question on the Vee Form.

Knowing Side

With the help of your teacher, your class will discuss the **Subject Area** background for this investigation. Then, with the help of your classmates, list the **Concepts** and the new **Vocabulary** words on your Vee Form. In the **Concept Statements** section of your Vee Form, use these words in sentences that define and explain them.

Procedure

You may wish to do this investigation as a demonstration, using a microprojector. Remind students not to put their hands in their mouths after touching any of the drugs.

1. With a medicine dropper, collect one *Daphnia* and squeeze it out of the dropper onto a clean depression slide. Add a coverslip and place it on the stage of a microscope.
2. Focus at low power and then locate the heart. Count the number of heartbeats for one minute, having your lab partner indicate when to start and when to stop counting.

Daphnia

3. Record the number of heartbeats in the graph in the **Additional Records and Observations** section on the back of your Vee Form. This is the control *Daphnia* heartbeat rate.
4. Place a small piece of paper towel next to the edge of one side of the coverslip. Add a drop of nicotine solution to the opposite edge, placing it half on and half off the coverslip. The paper will draw the solution across the slide.
5. Wait 15 seconds, then count the number of heartbeats as before and record in the graph on the back of your Vee Form.
6. Place the *Daphnia* in a 100-mL beaker that is half full of culture water. Clean the slide and coverslip thoroughly.

7. With a clean medicine dropper, collect a fresh *Daphnia* and place it on the slide. Add a coverslip and place it on the stage of a microscope. Focus again at low power.

8. Locate the heart and count the beats for one minute. Record this control rate on the graph as you did in step 3.

9. Add a drop of acetylcholine to the slide, using the same procedure used before.

10. Wait for 15 seconds, then count the heartbeats for one minute. Record your results on the graph.

11. Repeat this procedure until all the drugs have been tested. Remember to use a new *Daphnia* for each drug, and count the heartbeat rate before and after the drug is added. Record all results on the graph.

12. Display your results as a bar graph constructed from your data plotted on the back of your Vee Form.

13. Clean up your materials and wash your hands before leaving the lab.

In step 12, the students construct their bar graph on the <u>same</u> graph they used to plot their data.

Use the information on the **Knowing Side** of the Vee to interpret your results from the **Doing Side**, and then write your **Knowledge Claim**. Write a **Value Claim** for this study.

Analysis

1. Based on the results of your investigation, which of the drugs tested were stimulants and which were depressants?

Answers may vary. However, nicotine, adrenaline, coffee, and tea are known to be

stimulants. Known depressants include acetylcholine, aspirin, nasal spray, and alcohol.

_____

_____

2. Distinguish between stimulants and depressants on the basis of this investigation.

Stimulants increase the heartbeat rate and therefore metabolism. Depressants decrease

the metabolic rate.

_____

_____

Going Further

In the appropriate space on your Vee Form, write a **New Focus Question** that could be the point of a new investigation based on what you have just learned.

• Vary the concentration of the drugs to determine what effect this has on the heartbeat rate.

Name _____ Date _____ Class _____

How Drugs Affect the Heartbeat Rate

KNOWING SIDE

Focus Question:

What is the effect of some common drugs on heartbeat rate in *Daphnia*?

Subject Area:

Physiology
Pharmacology

New Focus Question:

What is the duration of each drug's effect on the heartbeat rate?

Concepts:

stimulant
depressant

Vocabulary:

acetylcholine

Concept Statements:

1. A depressant suppresses or slows down body metabolism.
2. A stimulant speeds up body metabolism.
3. Pharmacology is the study of the effects of drugs on the body.

DOING SIDE

Value Claim:

Knowing the effects of certain drugs on lower organisms can may provide information about their effects on humans.

Knowledge Claim:

The stimulants nicotine, coffee, and tea increase the heartbeat rate. The depressants acetylcholine, aspirin, nasal spray, and alcohol decrease the heartbeat rate.

Records:

[See Additional Records and Observations on the following page]

Materials:

10 medicine droppers, *Daphnia*, depression slides, coverslip, compound light microscope, paper towels, nicotine, watch with second hand, 100 mL beaker, acetylcholine, aspirin, antacid, cough drops, fresh perked coffee, nasal spray, 10% ethyl alcohol, fresh brewed tea

Procedure:

Determine the normal *Daphnia* heartbeat rate and compare this to the heartbeat rate under the influence of different drugs.

Additional Records and Observations:

Normal and Drug Induced *Daphnia* Heart Rate

Heart Rate (BPM)

700 — 600 — 500 — 400 — 300 — 200 — 100 — 50

control
nicotine
control
acetylcholine
control
aspirin
control
antacid
control
cough drops
control
coffee
control
nasal spray
control
10% alcohol
control
tea

Normal and Sickled Red Blood Cells

Introduction

Normal and sickle cell prepared slides are available from Ward's.
Human Blood
(93 M 6540)
Sickle Cell Anemia
(93 M 8120)

Human red blood cells normally have a shape that enables them to easily enter small capillaries. The hemoglobin molecules they carry deliver oxygen to the other body cells and then pick up carbon dioxide (CO_2). In some people, however, red blood cells (RBCs) take on a different shape, which is related to the hemoglobin's inefficiency in carrying oxygen.

The cause of sickling is a mutation in the series of amino acids that make up the hemoglobin in the red blood cells. This mutation occurs when the amino acid valine is substituted for glutamic acid in the chain of polypeptides that form the hemoglobin protein. The resulting red blood cells take on a sickle shape when the mutated hemoglobin loses its oxygen. When normal red blood cells lose their oxygen, they retain their shape.

Due to their shape, sickle cells tend to clog small blood vessels, causing intense pain. Sickle cell anemia occurs most often in persons of African descent.

Materials

- Red blood cell prepared slides
    normal
    sickle shaped
- Compound light microscope
- Colored pencils

Focus Question

Design your own **Focus Question** based on the information presented in the Introduction and Materials sections above, as well as the additional information presented in the Procedure section that follows. Write your Focus Question on the Vee Form.

Knowing Side

With the help of your teacher, your class will discuss the **Subject Area** background for this investigation. Then, with the help of your classmates, list the **Concepts** and the new **Vocabulary** words on your Vee Form. In the **Concept Statements** section of your Vee Form, use these words in sentences that define and explain them.

Procedure

1. Place the slide of normal red blood cells under the microscope and focus on low power. In the space provided in the **Additional Records and Observations** section on the back of your Vee Form, make a drawing of what you see.
2. Carefully switch your microscope to the high-power objective and focus on one red blood cell. In the space provided on the back of the Vee, make a drawing of what you see.
3. Repeat steps 1 and 2 with the sickle cell slide, again completing them by making drawings of the cells as they appear under low and high power.

Use the information on the **Knowing Side** of the Vee to interpret your results from the **Doing Side**, and then write your **Knowledge Claim**. Write a **Value Claim** for this study.

Analysis

1. What is the function of hemoglobin in red blood cells?

The hemoglobin carries oxygen.

2. Based on your reading in the text, would you say that sickle cell anemia is inherited as a dominant trait or as a recessive trait? Explain your answer.

Recessive; in defense of their answers, students could use a Punnett square to show how the disease is inherited.

3. Why is it important for red blood cells to be rounded in shape?

A rounded shape is necessary so that the cells can squeeze through small capillaries to supply body cells with oxygen.

4. Compare normal red blood cells with sickle cells.

Students should say something about the fact that normal red cells are rounded in shape with an indented center. Sickle cells are irregular and crescentlike in shape, leaving less room to carry oxygen and making them tend to clog up the small capillaries on their way to the body cells.

5. How might high altitudes affect a person with sickle cell anemia?

Since there is less oxygen in the air at high altitudes, the person with sickle cell anemia would have a very difficult time breathing. The shape of the blood cells in the person with sickle cell anemia makes it hard for the cells to carry sufficient oxygen, even at sea level.

Going Further

In the appropriate space on your Vee Form, write a **New Focus Question** that could be the point of a new investigation based on what you have just learned.

Before there was any medication to prevent malaria or to help people afflicted with malaria, it seemed that those people with sickle cell anemia were immune to malaria. It is possible that sickle cell anemia was a trait selected for in the face of an earlier outbreak of malaria. It may have evolved as a direct effect or a pleiotropic effect.

Name _____ Date _____ Class _____

Investigation 31-1
Normal and Sickled Red Blood Cells Vee Form Report Sheet

KNOWING SIDE

DOING SIDE

Focus Question:

What are the differences between normal red blood cells and sickled red blood cells?

Value Claim:

Understanding the differences between normal and sickled red blood cells gives researchers questions to investigate in the hope of curing sickle cell anemia.

New Focus Question:

Why has evolution made sickle cell anemia more common in persons of African descent?

Knowledge Claim:

Sickle cells are bent, not round like normal red blood cells. Round red blood cells are darker in the middle. The irregular shape of sickle cells makes it difficult for them to transport oxygen efficiently.

Records:

[See Additional Records and Observations on the following page]

Subject Area:

Cell theory
Blood functions

Concepts:

red blood cell
function
mutation

Vocabulary:

hemoglobin
sickle cell anemia

Materials:

prepared slides of normal and sickle-shaped human red blood cells, compound light microscope, colored pencils

Procedure:

Observe both red blood cell types, using both low and high power, then compare.

Concept Statements:

1. Hemoglobin on red blood cells carries oxygen.
2. Sickle cell anemia is an inherited disease in which the red blood cells are shaped so that they cannot carry enough oxygen.
3. Sickle cell anemia is caused by a mutation in the sequence of amino acids of hemoglobin.
4. Sickle cells clog capillaries and cause pain.
5. People of African descent are more likely to get sickle cell anemia.

Additional Records and Observations:

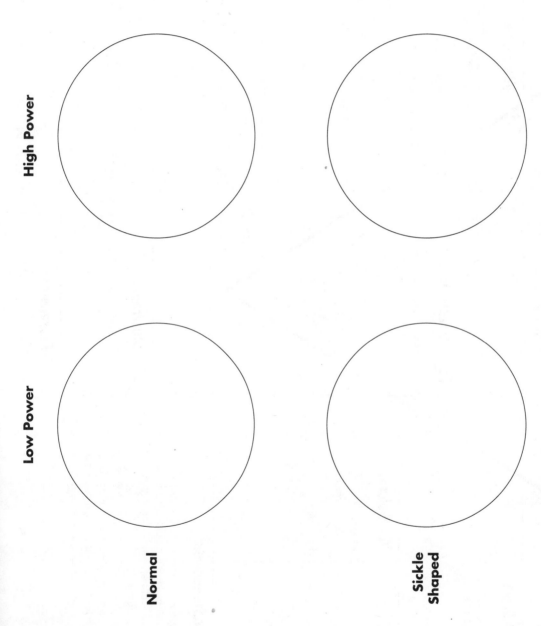

High Power

Low Power

Normal

Sickle
Shaped

HRW material copyrighted under notice appearing earlier in this work.

Disease Transmission Simulation

Introduction

During the school year, colds and flu spread from person to person, or an outbreak of chickenpox or other infectious disease may occur. All over the world, people are concerned about the transmission of disease. Most recently, the transmission of AIDS has become a great concern worldwide.

This investigation will involve the class in a simulation of disease transmission. Each of you will be given a dropper bottle of unknown solution and a clean test tube. Your personal dropper-bottle solution simulates aerosol droplets from a cough or sneeze, the bodily fluids which are exchanged in the transmission of AIDS, or those of any sexually transmitted disease. Since everyone will have clear liquids, the affected individual among you will have no "symptoms." After the simulation, you will try to identify the original infected person in the closed class population.

Materials

- Per student:
  "Unknown bodily fluid" solution
  Dropper bottle
  Test tube

- Phenol red indicator bottles

Focus Question

Design your own **Focus Question** based on the information presented in the Introduction and Materials sections above, as well as the additional information presented in the Procedure section that follows. Write your Focus Question on the Vee Form.

Knowing Side

With the help of your teacher, your class will discuss the **Subject Area** background for this investigation. Then, with the help of your classmates, list the **Concepts** and the new **Vocabulary** words on your Vee Form. In the **Concept Statements** section of your Vee Form, use these words in sentences that define and explain them.

Procedure

1. **CAUTION: Do not allow any solutions to touch your skin or clothing. Put on a lab apron, goggles, and disposable gloves.**
2. The liquids may look clear, but they are not simply water, so they should be handled with care. When your teacher says to begin, transfer 3 droppers full of your solution to your clean test tube.
3. Randomly pick one person. Let one partner pour the contents of their test tube into the other partner's test tube. Then pour half of the solution back into the first test tube. You two have now exchanged "bodily fluids" and you now share pathogens of any possibly transmittable disease either of you might have had. Record the name of your first exchange (Round 1) in the **Records** section of your Vee Form.
4. For Round 2, wait for your teacher's signal and then find a different student contact and exchange solutions in the same manner as in step 2. Be sure to record his/her name under "Round 2" on your **Records** section of the Vee. Do not exchange with any single person more than once. Repeat again for Round 3.

In preparing the solutions, the acidity of your tap water may sometimes cause it to test yellow and sometimes red with phenol red indicator. That's why dilute HCl is used instead. Students who become "infected" during the last round still will have sufficiently potent NaOH present, despite the dilution accompanied by the exchange.

Be sure to have the students rinse the test tubes well. Reserve the dropper bottles for the next class and you'll be ready to start again. If you fear students will not rinse them adequately and you want to prevent contamination in your next class, wash them yourself, or have a set of alternate tubes ready for the next class. The solutions retain their potency for at least a month.

Analysis

5. After all rounds are finished, your instructor will ask you to add one dropperful of phenol red to your test tube to see if your "bodily fluids" have become infected. Infected solutions will turn bright pink or red in color. All uninfected solutions will appear yellow. Record the outcome of your test in the **Records** section of your Vee.
6. If you are an infected person, give your name to your teacher. As names of infected people are placed on the chalkboard, record these in Table A in the **Additional Records and Observations** section on the back of your Vee Form.
7. Try to trace the original source of the infection, then determine the transmission route of the disease. First, cross out the names of all the uninfected persons from Table A. There should be only two people in Round 1 who were infected. One of these was the original carrier. Devise a method to determine the route of transmission, using the space to the right of Table A on the back of your Vee Form.
8. To test which person was the original carrier, pour a sample from his/her stock dropper bottle in a clean test tube and test with phenol red.

Who was the original disease carrier?

Answers will vary by class.

9. Clean up your materials and wash your hands before leaving the lab.

Use the information on the **Knowing Side** of the Vee to interpret your results from the **Doing Side**, and then write your **Knowledge Claim**. Write a **Value Claim** for this study.

1. After our three rounds, what was the number of students who were infected? Express this as a percentage of the number of students in the class.

Answers will vary depending on the number of students in your class.

2. What might be the number of infected individuals if we had exchanged fluids as many times in 3 minutes, as would happen if everyone sneezed every 30 seconds?

possibly 100% of the class

3. If there was an epidemic in your community, how would public health officials work to stop the spread of the disease?

They would post signs, alert the media, notify schools and other places of large public daily gatherings. These signs might ask people to stay at home if they show symptoms.

Going Further

In the appropriate space on your Vee Form, write a **New Focus Question** that could be the point of a new investigation based on what you have just learned.

Name _____ Date _____ Class _____

Disease Transmission Simulation

KNOWING SIDE

Subject Area:

Disease transmission

Concepts:

disease transmission

contact

Vocabulary:

epidemic

Concept Statements:

1. Transmission of disease may occur by exchange of bodily fluids, for example, in sneeze droplets and semen.

2. A contact is someone who has been close enough to another person to transmit or receive a transmittable disease.

Focus Question:

How can disease transmission be traced to the original carrier in a closed population?

New Focus Question:

How can a public health official distinguish between contact transmission and environmental transmission?

DOING SIDE

Value Claim:

In an epidemic, tracing an infected person would be important for controlling the spread of disease.

Knowledge Claim:

In a closed population, disease transmission can be traced to the original carrier by eliminating all uninfected contacts and examining the order of contacts.

Records:

Names of contacts:

Round 1 _____

Round 2 _____

Round 3 _____

Phenol red test

☐ red (positive)

☐ yellow (negative)

Materials:

"Unknown bodily fluid" solutions, dropper bottle, test tube, phenol red indicator bottles

Procedure:

Exchange "unknown bodily fluids" with 3 people, keeping record of your contacts in order of each round. Trace the disease to the original carrier.

Additional Records and Observations:

Answers will vary by class. This is given as an example of a trace.

Table A Infected Persons and Their Contacts in Three Rounds

Infected Person	Contacts		
	1	2	3
Kendra*	Jason	Mike*	Missy*
Missy*	Than*	Andra*	Kendra*
Than*	Missy*	Otis*	Andrew*
Greg*	Shayna	Jennifer	Doug*
Otis*	Dina	Than*	Greg*
Andra*	Elena	Missy*	Mike*
Mike*	Anne	Kendra*	Andra*
Andrew*	Earl	Dierdre	Than*

* Denotes infected person.

Transmission Route

Round 1	Round 2	Round 3

Missy ——→ Andra ——→ Mike

Missy ↑ Kendra

or ↕

Than ——→ Otis ——→ Greg

Than ↑ Andrew

Enzyme Action in Digestion

Introduction

See the annotation in the top margin on the next page to obtain Ward's materials for the preparation of solutions used in this lab.

Foods are composed of large organic molecules that are divided into three main categories: proteins, carbohydrates, and fats. None of these is water soluble, so each must be broken down into smaller units that can be used by the body. Proteins are broken down into amino acids, carbohydrates into glucose molecules, and fats into fatty acids and glycerol. Enzymes are the organic catalysts in the body that speed up the chemical breakdown of these molecules.

Egg white is almost pure protein. In this investigation you will be testing the action of two enzymes, pepsin and trypsin, on egg white. In addition, you will test the action of amylase (an enzyme found in saliva) on starch, and that of pancreatic lipase on fat.

Materials

Dispense all solutions from dropping bottles for safety.

Do **not** use salivary amylase in this lab. Use 1% amylase instead.

- Test tubes (10)
- Test tube rack
- Wax pencil
- Scalpel
- Egg, boiled
- Graduated cylinder, 10 mL
- Hydrochloric acid (HCl), 0.2%
- Pancreatin, 10%
- Sodium hydroxide (NaOH), 0.2%
- Litmus paper, blue
- Rubber stoppers (10)
- Starch solution, 1%
- Pepsin, 10%
- Lugol's iodine solution
- Amylase, 1%
- Beakers, 500 mL (2)
- Thermometer
- Hot plate
- Ice
- Soap solution
- Litmus paper, red
- Salad oil

Focus Question

Design your own **Focus Question** based on the information presented in the Introduction and Materials sections above, as well as the additional information presented in the Procedure section that follows. Write your Focus Question on the Vee Form.

Knowing Side

With the help of your teacher, your class will discuss the **Subject Area** background for this investigation. Then, with the help of your classmates, list the **Concepts** and the new **Vocabulary** words on your Vee Form. In the **Concept Statements** section of your Vee Form, use these words in sentences that define and explain them.

Procedure

Test A: Protein Digestion, Day 1

1. Place six test tubes in a test tube rack and label them A, B, C, D, E, and F.
2. **CAUTION: Scalpels are very sharp. Use extreme care.**

You may wish to cut up the egg white in advance to save class time.

3. With a scalpel, carefully cut some boiled egg white into about 5-mm cubes.

The following are available from Ward's.
Pancreatin (powder)
(30 M 2727) 20g
Pepsin (powder)
(39 M 2865) 25g
Amylase (powder)
(39 M 0058) 20g
Lugol's Iodine Stain
(39 M 1685) 500 mL

4. **CAUTION: Safety goggles, disposable gloves, and a lab apron should be worn for the remainder of this investigation.**
5. Add three cubes of egg white to each tube and then add the following:

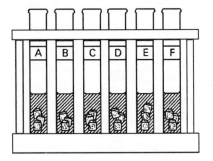

   * Tube A—10 mL of pepsin
   * Tube B—5 mL of pepsin and 5 mL of 0.2% HCl
   * Tube C—5 mL of pepsin and 5 mL of water
   * Tube D—10 mL of pancreatin (which contains trypsin)
   * Tube E—5 mL of pancreatin and 5 mL of 0.2% HCl
   * Tube F—5 mL of pancreatin and 5 mL of 0.2% NaOH

6. Test the solution in each tube with litmus paper and record the results in Table A in the **Additional Records and Observations** section on the back of your Vee Form.
7. Place a stopper in each of the test tubes and set them aside for one day.

Test A: Protein Digestion, Day 2

1. Retest each tube with litmus paper and record the results in Table A. Be sure to note any changes in pH.
2. Examine each of the tubes and note any changes in the size and consistency of the egg white. Record your observations in Table A of your Vee Form.

Test B: Carbohydrate Digestion

1. **CAUTION: Burns may result from careless handling of the hot plate. Be careful when using it to make your water baths.**
2. Obtain two clean test tubes and label them G and H. To each tube, add 5 mL of amylase solution, 5 mL of dilute starch solution, and two drops of Lugol's solution, then shake well. Note and record the color change that occurs in the tubes in Table B in the **Additional Records and Observations** section on the back of your Vee Form.
3. Fill a 500-mL beaker half full of water and place a thermometer in the beaker.
4. Place the beaker on a hot plate and heat the water to 37°C, and then maintain that temperature.
5. When the temperature is stable at 37°C, place tube G in the water bath.

Remind students to begin their water bath at the beginning of the experiment, so that it will be stabilized by the time it is needed.

Why is the water kept at 37°C?

Normal body temperature is 37°C. Enzymes are temperature-sensitive, and will denature at too high or too low a temperature. For example, enzyme denaturation may be fatal when someone has a fever of 106°F or higher.

5. Place tube H in a beaker filled with ice.
6. Allow both test tubes to remain undisturbed for 20 minutes.
7. Note any color change that occurs in the test tubes after 20 minutes, in Table B on the back of your Vee.

Test C: Fat Digestion

1. Obtain two more clean test tubes. Label the test tubes I and J. Place 10 mL of pancreatin solution, which contains pancreatic lipase, in tube I.
2. Add 10 mL of pancreatin solution and 4 mL of soap solution to tube J.
3. Add one drop of 0.2% NaOH to each tube. Shake each tube and test it with red litmus paper. Continue to add NaOH, one drop at a time, until both tubes become basic. Then add 4 mL of salad oil to each tube.
4. Place both tubes in a water bath at 37°C for 25 minutes.
5. Test each tube with blue litmus paper. Note and record any changes in Table C on the back of your Vee.
6. Clean up your materials and wash your hands before leaving the lab.

Use the information on the **Knowing Side** of the Vee to interpret your results from the **Doing Side**, and then write your **Knowledge Claim**. Write a **Value Claim** for this study.

Analysis

1. How do you account for any differences in reactions in the test tubes used in pepsin experiments?

   Pepsin works best in an acidic environment, such as that in the stomach. Therefore, the contents of tube B would be digested faster, since it contains HCl. The contents of tube A should be less thoroughly digested than those of tube B, and the contents of tube C would be even less digested than those of tube A.

2. What pH is most effective for trypsin activity? Where is trypsin produced in the body?

   Trypsin works best in a basic medium. It is produced by the pancreas, then released in pancreatic juice into the small intestine.

3. What effect does saliva have on starch? How could you prove this?

   Saliva contains amylase, which digests starch to form glucose. Tube G could be tested for the presence of glucose (using Benedict's or Fehling's solution test, or simple glucose test strips); if glucose is present, starch has been broken down.

4. How do you account for the results in the tubes containing the salad oil?

Soap emulsifies the oil, allowing for a faster rate of digestion of oil in tube J. When oil is

digested, acids are produced that cause the blue litmus paper to change to red.

_____

_____

_____

_____

5. From the results of this experiment, what factors influence the rate of enzyme action?

pH, temperature, and the size of particles being digested

_____

_____

Going Further

In the appropriate space on your Vee Form, write a **New Focus Question** that could be the point of a new investigation based on what you have just learned.

• Devise an experiment to test the optimum temperature or optimum pH at which digestive enzymes function.

Name _____ Date _____ Class _____

Enzyme Action in Digestion

KNOWING SIDE

Subject Area:

Nutrition
Physiology

Concepts:

fats
proteins
carbohydrates
emulsification

Vocabulary:

glycerol
fatty acids
amino acids
enzymes
catalysts

Concept Statements:

1. Foods are composed of large organic molecules divided into three main categories: fats, proteins, and carbohydrates.
2. No main food groups are water soluble.
3. Proteins are broken down into amino acids.
4. Carbohydrates are broken down into glucose molecules.
5. Fats are broken down into fatty acids and glycerol.
6. Enzymes are organic catalysts in the body that speed up chemical reactions, including those of digestion.
7. Emulsification is a physical breakdown of large fats into smaller pieces.

Focus Question:

What are the effects of the enzymes pepsin, trypsin, amylase, and pancreatic lipase on food molecules?

New Focus Question:

What is the optimum temperature and pH at which digestive enzymes function?

DOING SIDE

Value Claim:

Accessory organs such as the pancreas are important to the digestive system and are linked by way of a tube or duct.

Knowledge Claim:

Pepsin works best in an acidic environment, such as that of the stomach. Trypsin works best in a basic environment. Saliva has an enzyme called amylase that breaks down starch and forms glucose. Soap emulsifies oil, allowing for a faster rate of digestion of oil and the production of fatty acids.

Records:

[See Additional Records and Observations on the following page]

Materials:

test tubes and rack, wax pencil, scalpel, boiled egg, 10-mL graduated cylinder, 10% pepsin, 0.2% HCl, 10% pancreatin, 0.2% NaOH, blue and red litmus paper, rubber stoppers, 1% starch solution, Lugol's iodine solution, 1% amylase solution, 500-mL beakers, thermometer, hot plate, ice, soap solution, salad oil

Procedure:

A: Test effects of pepsin and trypsin on the digestion of egg white in acidic, neutral, and basic conditions overnight. B: Test the effect of amylase solution on starch digestion at both 37°C and at 0°C for 20 minutes. C: Test the effect of pancreatic lipase on the digestion of salad oil, with and without the aid of a soap solution.

Additional Records and Observations:

Table A Protein Digestion

Tube	Day 1 pH	Day 2 pH	Day 2 Observations
A. 10 mL pepsin			less digested than tube B
B. 5 mL pepsin 5 mL 0.2% HCl			more is digested
C. 5 mL pepsin 5 mL H$_2$O			less digested than tube B
D. 10 mL pancreatin			less digested than tube F
E. 5 mL pancreatin 5 mL 0.2% HCl			less digested than tube B
F. 5 mL pancreatin 5 mL 0.2% NaOH			most digested

Table B Carbohydrate Digestion

Tube	Initial Color	Color After 20 Minutes in:
G	blue	hot water bath: iodine color
H	blue	ice bath: blue

Table C Fat Digestion

Tube	pH Pretest	pH Test After 25 Minutes in Hot Water Bath
I		not as acidic
J		acidic; changes blue litmus paper to red

Lactose Digestion

Introduction

When the disaccharide lactose is properly digested it is split into the monosaccharides glucose and galactose by the enzyme lactase. In some humans, lactase activity declines to low levels in late childhood or adult life. This results in an inability to digest lactose. When this happens, much of the lactose is used by bacteria in the large intestine to produce certain acids that tend to increase intestinal motility (spontaneous motion), causing gas, bloating, and diarrhea. Lactose intolerance is found in 20 percent of Caucasians and 80 percent of persons of Asian or African descent. People lacking this enzyme can be helped by using a product that enables them to digest dairy products.

In this investigation you will design a control experiment to test the effectiveness of a liquid milk-treatment product designed to enable lactose intolerant people to drink milk. You will also attempt to find out how it works.

Materials

- Milk-treatment product (liquid)
- Toothpicks
- Depression slides
- Reference books
- Glucose test paper
- Glucose solution
- Milk
- Medicine droppers
- Benedict's solution and/or other suitable test solutions
- Hot plates, water, beakers, test tubes, etc.

Focus Question

Design your own **Focus Question** based on the information presented in the Introduction and Materials sections above, as well as the additional information presented in the Procedure section that follows. Write your Focus Question on the Vee Form.

Knowing Side

With the help of your teacher, your class will discuss the **Subject Area** background for this investigation. Then, with the help of your classmates, list the **Concepts** and the new **Vocabulary** words on your Vee Form. In the **Concept Statements** section of your Vee Form, use these words in sentences that define and explain them.

Procedure

Bring some biochemistry reference books and physiology texts to class that the students can peruse to supplement their textbook.

After an acceptable procedure is discussed among the student team, make sure their test measures are adequate for a depression slide or test tube.

This is a student-designed investigation that requires critical-thinking skills and creativity in order to design their own experimental procedure.

1. Read the container labels for the milk-treatment product so you know what it is and what it does. Remember, your task is to find out how it works.
2. Design a control experiment that provides evidence that the product does what it claims to do. Pretend you are a member of a division of the U.S. Food and Drug Administration research lab that is testing this product for the public market. Feel free to browse through some of the reference materials your teacher may have brought to the lab.
3. Your group will design and write a procedure for your experiment at the base of your Vee Form. Be sure to include the materials from the list above that you choose to use and ask the teacher for others that may not be listed.
4. After you have proposed a procedure and have decided on materials to use, consult your teacher. The procedure must be discussed and approved by your teacher before you may proceed with the actual testing.

If students use Benedict's solution, or any other skin irritant, be sure they wear apron, goggles, and gloves.

5. Decide on the format of your data table(s), and design them on the **Records** or **Additional Records and Observations** section of your Vee Form. Record all data in these tables.

6. Clean up your materials and wash your hands before leaving the lab.

Summarize the results of your experiments in tables, charts, graphs, or descriptive observations in the Records and Additional Records sections of your Vee Form.

Use the information on the **Knowing Side** of the Vee to interpret your results from the **Doing Side**, and then write your **Knowledge Claim**. Write a **Value Claim** for this study.

Analysis

If students want to test a treated quart of milk with Benedict's test, change their procedure steps to adding a drop of treated milk to a drop of Lactaid®, and the use of a glucose test strip to save time. If you aren't concerned about the time, but you are concerned about the cost of milk, allow them to do the Benedict's test with only 10 mL of milk in a test tube.

1. How did you test the reliability of the glucose test paper, of the Benedict's solution test, or of any other test you used?

   Answers will vary. For example, testing a known glucose solution to get a baseline reading.

2. How did you feel about designing your own investigation and seeing it through?

   Answers will vary.

3. Does the product do what it says it does? What does it do?

   Yes; it breaks down lactose.

Going Further

In the appropriate space on your Vee Form write a **New Focus Question** that could be the point of a new investigation based on what you have just learned.

The active ingredient in Lactaid® is lactase; when this is placed in milk, digestion of the lactose occurs readily.

Lactaid® is distributed by McNeil Consumer Products, 7050 Camp Hill Rd., Fort Washington, PA 19034. You can call 1-800-522-8243 to place an order or to request a free pamphlet on lactose intolerance. An information sheet is sent with each order.

In preparation for this lab, buy Lactaid® liquid from a local pharmacy or grocery store. If students use depression slides for this investigation, they will use such small amounts that the supply will last for years.

Do not use the tablet form of Lactaid®, since it tests positive for glucose.

Be sure you read the information sheet accompanying the Lactaid®. You may choose to read parts of it to your students before they make their lab design, but refrain from reading the section "How Lactaid® Works." They determine this themselves in lab.

Name _____ Date _____ Class _____

Lactose Digestion

KNOWING SIDE

DOING SIDE

Focus Question:

What is the biochemistry of lactose? What is the chemistry of lactase enzymes? Does Lactaid® do what it is claimed to be able to do in a student-designed experiment?

Value Claim:

The U.S. Food and Drug Administration research lab designs studies like this to test a multitude of products for human consumption.

Subject Area:

Digestion
Physiology

New Focus Question:

Would blood sugar level rise after a milk-intolerant person drinks milk with Lactaid®?

Knowledge Claim:

Lactaid® has an enzyme that breaks down lactose into glucose and galactose. Lactaid® digests lactose in milk, so it works as it claims it does.

Records:

Observation Data:

Vocabulary:

enzyme
Lactaid®
lactase
lactose
glucose
galactose
disaccharide
monosaccharide

Concepts:

milk intolerance
research
control

Materials:

Lactaid® liquid, toothpicks, depression slides, reference books, glucose test paper, glucose solution, milk, medicine droppers, suitable test solutions, other materials according to the procedure selected.

Procedure:

Answers will vary. Example: Test for presence of glucose in a given glucose solution, milk, Lactaid®, and a Lactaid®-milk solution.

Concept Statements:

1. Lactaid® contains yeast-derived lactase for people who are milk-intolerant.
2. A research study is directed by questions based on what is already known and uses a control for comparison with the experimental situation.
3. Enzymes are important in the buildup and breakdown of chemicals in the body.
4. Milk-intolerant people do not produce the enzyme lactase.
5. Lactose is a disaccharide; glucose and galactose are monosaccharides.

Additional Records and Observations:

Answers will vary.
Possible data table:

Test for Glucose	Result	Glucose Present?
Milk	–	no
Lactaid®	–	no
Lactaid® + Milk	+	yes
Glucose	+	yes

Embryonic Development

Introduction

Slides of sea stars or sand dollars, for example, may be substituted for the sea urchin slides, since early development is very similar among echinoderms.

Materials

Focus Question

Knowing Side

Procedure

The prepared slides needed in this lab are available from Ward's.

Sea Urchin Embryology
(92 M 8330) composite

Chick 28-hour
(92 M 9035)

Chick 33-hour
(92 M 9040)

Chick 43-hour
(92 M 9058)

Chick 56-hour
(92 M 9070)

Most multicellular organisms begin life as a single cell—the fertilized egg, or zygote. The fertilized egg divides many times, the new cells then begin to specialize as they become part of specific tissues, and complex structures are formed as the embryo grows into a fully developed organism. The ways in which different species carry out these essential steps during embryonic development takes place reflect both the life cycle of the species and the evolutionary relationships among species.

- Microscope, compound light
- Stereomicroscope
- Prepared slides:
  Sea urchin embryos,
  mixed developmental stages

- Chicken embryos,
  whole mounts,
  28-hour, 33-hour,
  43-hour, 56-hour

Design your own **Focus Question** based on the information presented in the Introduction and Materials sections above, as well as the additional information presented in the Procedure section that follows. Write your Focus Question on the Vee Form.

With the help of your teacher, your class will discuss the **Subject Area** background for this investigation. Then, with the help of your classmates, list the **Concepts** and the new **Vocabulary** words on your Vee Form. In the **Concept Statements** section of your Vee Form, use these words in sentences that define and explain them.

1. Place the slide of sea urchin embryos on the stage of the microscope and locate an embryo, using low power. Center the embryo, and then examine it under high power. Examine several embryos until you have seen the following stages of sea urchin development: 2 cells, 4 cells, a hollow ball of many cells (blastula), a hollow ball with a central column of cells (gastrula), and a triangle–shaped larva. Make a drawing of each of the five stages in the **Additional Records and Observations** section on the back of your Vee Form.

Yolk is stored in eggs to provide nourishment for the developing organism until it is able to obtain food from its environment. Sea urchins develop from a fertilized egg to a feeding larva in only 48 hours.

How much yolk do you think is stored in a sea urchin egg? Why?

Very little. It does not require much stored food because it develops quickly into a small larva.

_____

_____

In species whose eggs must contain large amounts of yolk, the egg cell is large, with most of the yolk stored in the bottom of the egg. It is difficult for the egg cell to divide through these large yolky regions.

How does the amount of yolk in a sea urchin egg affect the way in which it divides?

It divides easily into four equal size cells because there is little yolk to interfere with the expansion that accompanies cell division.

If students wish to view chick embryo whole mounts with the compound light microscope, caution them to use the high power objective very carefully. These slides are usually thick and can be broken very easily by the high-power objective if it is lowered too far while focusing.

2. Chicken eggs must be incubated about three weeks before the chick hatches. Bird eggs store such great quantities of yolk that only a very small amount of active cytoplasm near the top of the egg is able to divide. The rapidly dividing cells form a flattened disc about 3 mm in diameter that then gives rise to the differentiated tissues of the embryo. Examine the whole mounts of the chicken embryos using the stereomicroscope. At this stage of chicken development, the chicken embryo is growing rapidly, using the stored yolk to supply nutrients. Complex organs such as the brain and heart are beginning to form. Compare your slides with the diagrams of two different stages of chicken embryos below.

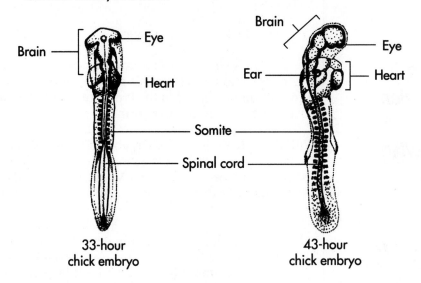

33-hour chick embryo

43-hour chick embryo

At what stage does the heart appear?

33-hour

Describe how the shape of the heart changes as it develops.

It forms a tube that bulges to the side as it grows, then it folds to form a loop.

How do the eyes begin?

as hollow protrusions of the brain

Blocks or segments of tissue called somites give rise to bones and muscles in vertebrates, and display the evolutionary relationship between vertebrates and other segmented organisms such as segmented worms and arthropods.

How many pairs of somites does each embryo have?

~~Answers may vary slightly, but should be close to 8 somites at 28 hours, 12 somites at 33~~ hours, 19 somites at 43 hours, and 29 somites at 56 hours.

3. Describe or draw your observations in the **Additional Records and Observations** section on the back of your Vee Form.
4. The earliest stages of the embryos of birds, reptiles, and mammals are remarkably similar. Each forms a flattened disc of cells like the chicken embryo and begins to form the major organ systems characteristic of vertebrates in much the same way. As the embryos develop further, they acquire more and more of the characteristics that are unique to their species. Look at the drawings of the bird, reptile, and human embryos below.

(a)   (b)   (c)

Which drawing is the human embryo?

~~The first embryo is a turtle, the second one a chicken, and the third one a human. Students~~ may or may not recognize the human embryo, but should notice the similarities among the ~~three.~~

5. Clean up your materials and wash your hands before leaving the lab.

Use the information on the **Knowing Side** of the Vee to interpret your results from the **Doing Side,** and then write your **Knowledge Claim.** Write a **Value Claim** for this study.

Analysis

1. How does a developing human obtain nourishment before birth?

from the mother via the placenta

2. How much yolk would you expect to find in a human ovum? Why?

Very little. It does not require much stored food, since it begins to obtain nourishment from the mother soon after fertilization.

3. Would you expect a human egg to divide like the sea urchin, or the chick? Why?

Sea urchin. The small amount of yolk allows it to divide completely into equal size cells.

4. Why might it be surprising that both the chicken embryo and the human embryo first form a flattened disc of cells before gastrulation?

because the human embryo is not surrounded by a large quantity of stored yolk like the embryos of birds and reptiles

5. Why do you think the human embryo develops its basic structures from a flattened disc rather than from a hollow ball like the sea urchin embryo?

because it is more closely related to bird embryos than those of sea urchins and initially sets up its body plan and develops organs in a similar way

6. Do any parts of the fully developed human body reflect the early segmentation seen in the somites of the embryo?

Vertebrae and ribs develop directly from the segmented somites, and major blood vessels and nerves also follow the pattern of the somites. Muscles also reflect segmentation, but less clearly in the fully developed human.

7. Why would an embryologist study chicken embryos in order to find out how to prevent birth defects in humans?

Severe birth defects usually arise during the early stages of embryonic development. During these stages, birds and humans follow similar developmental patterns.

Going Further

In the appropriate space on your Vee Form, write a **New Focus Question** that could be the point of a new investigation based on what you have just learned.
- Collect frog eggs from a nearby pond or creek and rear them in an aquarium, or obtain prepared slides of frog embryos at various stages of development. Frog eggs contain much less yolk than bird eggs, but most have much more yolk than sea urchins. Compare the development of the frog with that of the sea urchin and the chick, and relate the developmental patterns of the frog to its life cycle.

Name _____ Date _____ Class _____

Embryonic Development

KNOWING SIDE

Subject Area:
Embryology
Growth and development

Concepts:
development
differentiation

Vocabulary:
zygote
embryo
blastula
gastrula
somites

Focus Question:
How does embryonic development progress in invertebrate (sea urchin) and vertebrate (chicken) embryos?

New Focus Question:
How does development progress in other animals, such as amphibians, fish, or reptiles?

Concept Statements:
1. A zygote, or fertilized egg, divides and differentiates to form an embryo, which continues to grow and differentiate into a fully developed organism.
2. At the early stages of cellular division, the dividing zygote forms a blastula, or hollow ball of many cells, and then a gastrula, or hollow ball with a central column of many cells.
3. Somites are blocks or segments of tissue that give rise to bones and muscles in vertebrates, and the body segments of segmented worms and arthropods.

DOING SIDE

Value Claim:
The study of embryonic development of sea urchins, and of toxic substances that impair this development, can aid in the study and prevention of human birth defects.

Knowledge Claim:
Development progresses gradually through a series of cellular divisions and differentiation. The amount of yolk in the egg affects the development of the embryo, and is determined by need for nutrition and influenced by evolutionary relationships.

Records:
[See Additional Records and Observations on the following page]

Materials:
compound light microscope, stereomicroscope, prepared slides of sea urchin embryos (different stages), chicken embryos (different stages).

Procedure:
Observe and compare the various stages of embryonic development of sea urchins and chickens.

Additional Records and Observations:

Descriptions of Chicken Embryos:

28-hour

33-hour

43-hour

56-hour

2 cell

4 cell

blastula

gastrula

larva